Humanities As Science Matters

History, Philosophy & Arts

Editors: Maria Burguete & Hauke Riesch

The fourth biennial Science Matters (Scimat) conference "Humanities as Science Matters: History, Philosophy and Arts" was held at the emblematic Casa do Infante, Porto, Portugal, October 15-17, 2013.

Published by Pantaneto Press
Luton, UK

ISBN 978-0-9929411-2-3

Printed and bound by CPI Group (UK) Ltd, Croydon, CR0 4YY

Contents

Contributors

Paulo Bernardino Bastos (MA, PhD) is director of the Post-Graduate Program (Master) in “Contemporary Artistic Creation”, in the Department of Communication and Art at the University of Aveiro (Portugal), where he teaches. Researcher at iD+ (Research Institute for Design, Media e Culture). pbernard@ua.pt

Maria Burguete got her PhD in History of Science (contemporary chemistry) from Ludwig Maximilians University at Munich, Germany (2000). She is a scientist and a university lecturer with teaching and research experience in a wide variety of scientific fields. She is a scientific researcher at Scientific Research Institute Bento da Rocha Cabral in Portugal since 2007. She has published 11 books in the scientific field (and seven books in poetry) and over 25 scientific papers. She is a corresponding member of European Academy of Sciences, Arts & Letters since 2010. mariamartins434@gmail.com

Jean-Patrick Connerade is of French nationality, born in London in 1943. He is an Emeritus Professor in the Quantum Optics and Laser Science Group at the Physics Department, Imperial College London, and was educated at the Lycée Charles de Gaulle, then State Scholar at Imperial College (University of London) where he obtained his degrees up to DSc level in Physics. He has worked at Culham, the European Space Agency, Universität Bonn, the Ecole Normale Supérieure, Wuhan Institute of Physics and Mathematics, East China University, etc. and held the Lockyer Chair of Physics in the University of London for many years. Currently, he is the President of the European Academy of Sciences Arts and Letters, Honorary President of Euroscience, President Jury of the Rammal Medal and Honorary Professor of Physics at the East

China University of Shanghai, Permanent Visiting Researcher at the Wuhan Institute of Physics and Mathematics of the Chinese Academy of Science. Amongst Professional Societies he belongs to are the Société des Poètes Français, the Société des Gens de Lettres, the World Academy of Science and Art and he is a Fellow of the Institute of Physics and the Royal Society of Chemistry. He is the author of the Monograph 'Highly Excited Atoms' (500 pages) published by the Cambridge University Press 1998 (translated into Chinese and published CAS Press Beijing 2003) ISBN 7-03-010814-0 reprinted CUP paperback. Co-author and Co-editor of many books with World Scientific Press, Plenum Press, etc. His Literary works (in French) include books of prose and poetry published by the Editions l'Age d'Homme (Paris and Lausanne) the Editions des Poètes Français 5Paris) under the pseudonym 'Chaunes' and can be found on www.amazon.fr. His Literary Prizes include the Prix Jose-Maria de Heredia de l'Académie Française, the Prix Paul Verlaine de la Maison de Poésie de Paris and the Grand Prix Victor Hugo de la Société des Poètes Français. jean-patrick@connerade.com

Sérgio Eliseu (MA, Ph.D. student). BA in Art History from F.L.U.C. (University of Coimbra) and a Master Degree in Contemporary Artistic Creation (University of Aveiro). Currently he is studying towards an Art and Design PhD, at the FBAUP (University of Porto) with a scholarship by FCT (Foundation for Science and Technology). Researcher at I2ADS (Institute of Research in Art, Design and Society) and iD+ (Research Institute for Design, Media e Culture). sergio@eliseu.com

Anne-Sophie Godfroy obtained her PhD in 2004 at University Paris-Sorbonne, after studying Philosophy at the Ecole Normale Supérieure in Paris. She is associate professor at University Paris-Est-Creteil and researcher at the research centre "Sciences Normes Décision" (Université Paris-Sorbonne & CNRS, FRE 3593). Her main research interests are methodology for international comparisons, science in society, gender. For ten years, she participates to several European research projects on gender and Science and Technology. anne-sophie.godfroy@paris-sorbonne.fr

Lui Lam, humanist and physicist, obtained his BS (First Class Honors) from University of Hong Kong, MS from University of British Columbia, and PhD from Columbia University. He is a physics professor at San Jose State University, California, and adjunct professor at Chinese Academy of Sciences and the China Association for Science and Technology. Lam invented Bowlics (1982), one of three existing types of liquid crystals in the world; Active Walks (1992), a new paradigm in complex systems; and two new disciplines: Histophysics (2002) and Scimat (Science Matters, 2007/2008). He published 16 books and over 180 scientific papers. He is the founder of the International Liquid Crystal Society (1990); cofounder of the Chinese Liquid Crystal Society (1980); founder and editor of two book series: Science Matters (World Scientific) and Partially Ordered Systems (Springer). His current research is in scimat, histophysics and complex systems. Email: lui2002lam@yahoo.com

Maria Manuela Lopes is a visual artist and researcher who works in Portugal and the United Kingdom. Her current work is interdisciplinary and scientifically informed, researching the relationship between memory and self-identity and presented to the public through multimedia installations which occasionally include biological materials. Her work has been exhibited nationally and internationally and the artist has been teaching in Portugal since 1998. She studied sculpture at FBAUP – Oporto and holds an MA from Goldsmiths College in London. She gained a practice-based PhD at UCA – Farnham, UK (under the supervision of Kathleen Rogers, João Lobo Antunes, and Judith Williamson) in the area of New Media-Fine Arts. Lopes is currently beginning a Post Doc research project expanding on her previous research findings and located within a molecular biology laboratory, botanical gardens and museums, and neuroscience clinical and pathology contexts. She co-directs two artistic residency programs in Portugal

Ectopia and Cultivamos Cultura. She has presented her projects at conferences as well as having been published both nationally and internationally. maria@manuelalopes.com

Manuela Monteiro is an anatomopathology technician at the Newcastle Brain Tissue Resource, Institute of Ageing and Health, Newcastle University. mmapgfm81@gmail.com

Filipa M. Ribeiro is PhD Fellow at the Faculty of Psychology and Educational Sciences of University of Porto and visiting researcher at Egolab at the Department of Social and Cultural Anthropology at the Autonomous University of Barcelona. filipa.ribeiro@gmail.com

Hauke Riesch is a lecturer in sociology at Brunel University London, where he researches on the public understanding of science, philosophy of science, environmental and energy policy. As a member of the newly created Brunel Centre for Comedy Studies Research, he has decided to do something fun for a change and look at the links between humour and science. hauke.riesch@brunel.ac.uk

Nigel Sanitt obtained his BSc from Imperial College London and Part III of the Mathematics Tripos and PhD from Cambridge University. He trained as an astrophysicist at the Institute of Astronomy, Cambridge, before becoming a lecturer at University College, London University. He is founder and editor of The Pantaneto Forum. The journal aims to promote debate on how scientists communicate, with particular emphasis on how such communication and research skills can be improved through a better philosophical understanding of science. nigel@pantaneto.co.uk

David Schmool gained a BS (Hons.) Degree in Physics at the University of Portsmouth, UK and D. Phil. at the Department of Physics, University of York, UK. After completing doctoral studies he held research posts at the University of Liverpool, UK; Istituto MASPEC CNR, Parma, Italy; University of the Basque Country, Spain; University of Versailles,

France and the University of Exeter, UK. He has been visiting scientist at the following institutes: Simon Fraser University, Canada; the University of Glasgow, UK; the University of Perpignan, France, the University of Versailles, France and the University of Duisburg-Essen, Germany. He has recently written a book on Solid State Physics and is the author / co-author over 200 refereed publications including 7 book chapters and over 60 research papers. Until recently he was lecturer in Physics in the Department of Physics and Astronomy at the University of Porto, Portugal. He is group leader for the Ultrafast Laser and Magnetodynamic Spectroscopies unit of the Institute of Physics of Materials of the University of Porto – Institute of Nanotechnologies (IFIMUP – IN) and also Vice-President of the Institute. He is currently Professor at the University of Perpignan (Université de Perpignan Via Domitia), France and researcher at the Laboratoire PROMES, CNRS, Perpignan. His principal research interests include magnetism and magnetic materials, nanotechnology, nanomagnetism, spin dynamics in nanosystems, ultrafast processes in magnetism and solid state materials, optical properties of nanomaterials. dschmool@fc.up.pt

Foreword

For at least a generation, there has been a prejudice that *specialisation* was somehow connected with *quality*. Thus, for example, one might choose to be a historian or a mathematician, but a historian of mathematics would be diluting both fields and therefore was considered to be achieving less. This prejudice became so strong that even well-established areas, such as physics, began to fragment internally, developing their own hierarchy of importance, with high-energy physics at the apex. This led to the development of so-called 'sub-cultures', characterised by a propensity to reinvent the wheel. For example, an atomic physicist might honestly believe he had discovered a new effect or even a new theory, whereas in fact it would be well-known and common knowledge in, say, nuclear physics. Because they had become perceived as 'different' subjects, with separate communities and separate journals, specialisation (with its implicit ignorance of fields outside) would lead to claims of novelty when in fact there was none.

The prejudice somehow seeped through into university education. A broad course, leading to a 'general' degree, was par less prestigious than a more detailed and particular course, leading to a 'special' degree. The hierarchy of terms used seemed in conflict with the very origin of the word: *university*, to which few people gave any thought.

In part, it must be admitted, the prejudice was founded on a healthy dose of humility. The idea was that humanity had progressed as far as it could in the combined understanding of many disciplines. It would be arrogant, therefore, to continue along this path. To avoid being out of one's depth, one should not even attempt to learn more than one discipline at a time. Indeed, for fear of being taxed with arrogance, and in the interest of not wasting the most productive years of their lives, intellectuals should stop studying at a certain age and, from then on, concentrate on the one subject they had mastered and become as creative as possible within its boundaries (a process known as *research*).

For a time, this approach to culture replicated socio-economic development. Indeed, society was believed to work better if each individual practised only his own well-defined skill, rather as on an assembly line, where every worker performs his particular task, eventually aspiring to become the world's expert on a single gesture, many times repeated and constantly perfected.

With the realisation that specialisation was encountering limits, however, came the notions of *multidisciplinarity* and of *interdisciplinary areas*. These ideas, in principle quite sensible, nonetheless remain static, because they rest on the postulate that boundaries persist and do not evolve. A perfect example of the difficulty this leads to was the futile and inconclusive debate on whether one supposedly new area should be termed 'physical chemistry' or 'chemical physics', brought to a close by Richard Feynman when he declared that he did not understand the difference between physics and chemistry. Against this, biophysics or biochemistry can be regarded as a success story, because the whole proved to be more than the sum of the parts and also led to technological innovations.

Today, however, the form of 'assembly-line' industrialisation outlined above has run its course. As we are all aware, in the modern economy, a single individual will have to adapt several times over to innovation and change by training and retraining in different and sometimes seemingly unrelated fields, broadening his experience and deepening his knowledge. We have suddenly discovered that society is *complex*, by which we mean that every part depends on every other part (the very *opposite* of specialisation). Those with broad knowledge and experience, in this new environment, will clearly have the advantage. The fact is that further specialisation, in the modern world, no longer leads anywhere. It is, as one says, dead in the water today.

And so, perhaps in harmony with this fundamental revolution, comes the re-discovery of complexity in the world of contemporary knowledge and culture. The idea that culture is a whole, that it is wrong to consider it merely as an assembly of separate parts, is not, of course, a new one. It is, in fact, the very basis of what used to be called 'humanism', the starting point of post-mediaeval civilisation, from which, somehow, we managed to diverge into the blind alleys of excessive specialisation. So, we may regard the intellectual revolution of today, the end of specialised thinking, as rediscovering the importance of the whole.

Of course, it would still be unrealistic to advocate becoming an *absolute* generalist. Like all ideals, ever-broadening one's knowledge is a permanent and lifelong challenge. However, in today's perception, it is no longer interpreted as a sign of arrogance. Rather, it is a question of *survival.* This applies both to the individual and to the subject concerned. Science cannot continue along the road towards further specialisation without becoming irrelevant or sinking into ignorance. Likewise, artists cannot just be artists without painting themselves off into a corner of the modern world.

Whatever social, economic or political system one may believe in ceases to be the most important question today. The key issue is how to be sufficiently aware of others – other fields, other cultures, other people – to spot new opportunities when and where they arise. A common culture is therefore essential to modern man. That is precisely what the *scimat* movement seeks to achieve: it consists in developing this new awareness and the adaptability of thinking which will open up new opportunities in an ever-changing world.

The ambitious aims of the scimat project will obviously not be achieved at a single stroke. Nor can they. We are describing the start of a process, and the present book, which brings together a wide range of contributions from widely different fields, presented at a discussion meeting of the scimat group in O' Porto, is merely a step along the way. The idea is not simply to place these different fields alongside each other in the style known as eclecticism. The purpose is rather to place them in a common context and see how they may be related enough to influence each other.

We do not claim to have achieved the revolution we describe. We do claim, however, that the process initiated by the scimat project is essential to humanity, and that real progress cannot be made without it.

Jean-Patrick Connerade
European Academy of Sciences Arts & Literature, Paris

Preface

The fourth biennial Science Matters (Scimat) conference "Humanities as Science Matters: History, Philosophy and Arts" was held at the emblematic Casa do Infante, Porto, Portugal, October 15-17, 2013. Scimat is a multidiscipline that focuses on raising the scientific level (or *scientificity*) of the humanities, by encouraging interaction and collaboration between humanists and natural scientists.

Most articles in this book were presented at the conference. The interdisciplinarity is well portrayed in this volume where we find four main areas: The **History of Science** with the contribution of Anne-Sophie Godfroy where "A History of the History of Gender & Science" is presented in chapter 8. Afterwards, in chapter 7 David Schmool refers to the "Insights into Solid State Physics" where examples of the relation between the internal structure of solids and physical properties of materials are given. Chapter 4, by Maria Burguete, describes further developments of an historical description of medical sciences involving an emblematic case study in Portugal with international repercussions - The Generation of 1911.

The **Philosophy of Science** with two chapters: 1 and 9. Chapter 1, "The Scimat Program: Towards a Better Humanity" by Burguete and Lam, recounts the basic tenets of scimat (proposed by Lam) and the development of the international effort and action plan in promoting it since 2007. Scimat´s key 1-2-3 insight is that we have "one culture, two systems, three levels", i.e., "science culture, simple and complex systems, three research levels". This idea is further reinforced in chapter 9, "Philosophy, Science and Scimat" by Lui Lam.

The **Arts** are well represented in this book where we have new insights of "Visual Arts Research Practice in Scientific Laboratories" given by Maria Manuela Lopes in chapter 6 and also in chapter 5 together with her colleagues Sergio Eliseu, Paulo Bastos and Manuela Monteiro, showing the "Exploratory Reality Art Based Projects in Anatomic Pathology Laboratories". The answer to the question: how different is fiction writing from science writing? is given by Nigel Sanitt. This scientist began a new adventure as a novelist writing his first novel, inspiring him to the title of chapter 11: "From Science to Fiction".

The **Sociology of Science** is awarded the contribution of Jean-Paul Connerade with his chapter "Science, Culture and 'Anti-science'" in chapter 3. The thematic of education where the strength of higher education values are exposed and discussed by Filipa Ribeiro on "Combining Permanence and Change in Higher Education: The Strength of Strong Values" (Chapter 2). The sociologist Hauke Riesch on chapter 10 gives a brief account on humour appearing in serious scientific publications in his article "Humour, Science and Seriousness".

We believe a new era of scientific knowledge is emerging with a new and unified perspective of the Humanities (including History, Philosophy and Arts) in a unique multidiscipline called Scimat.

1

The Scimat Program: Towards a Better Humanity

Maria Burguete and Lui Lam

Scimat (Science Matters) is a new term coined by Lui Lam in 2007/2008. *Conceptually*, Scimat represents the four tenets that (1) Science is humans' effort to understand Nature without bringing in God or any supernatural; (2) Science covers everything in Nature; (3) Nature includes humans and all nonhuman systems; (4) all research on human matters, humanities in particular, are part of science. *Disciplinarily*, Scimat represents the collection of research disciplines that deal with humans; thus, Scimat is the sum of the humanities, social science and medical science. In particular, scimat aims to raise the scientific level of the humanities which, *not* "natural science", cover the arts and decision making that determine the well-being of the human race. Thus, the ultimate aim of scimat is not just for a unified knowledge, but for a better humanity. Towards this end, The Scimat Program was started by Maria Burguete and Lui Lam in 2007, which is the latest international effort to revive the Aristotelian tradition of a unified knowledge. It has (1) established a biennial international scimat conference series; (2) formed an international committee (ISMC) to advocate the scimat concept and oversee scimat activities; (3) established a Science Matters Series of books (World Scientific). The Scimat Program is in the midst of establishing a number of scimat centres around the world. Additionally, a scimat general-education course for students of any major and any level is under construction and test-teaching. In this chapter, the motivation, significance and progress of The Scimat Program are outlined.

1.1 Introduction

The tradition of treating everything in the universe (consisting of human and non-human systems) from a unified perspective—starting with Aristotle and continued until the Renaissance—was broken with the rise of modern science in the last 400 years. Concerted efforts to revive this tradition happened from time to time and failed, except that the Enlightenment succeeded partially (see Section 1.4).

The Scimat Program is the latest international effort to revive the Aristotelian tradition of a unified perspective of knowledge that treats research on humans and nonhuman systems as equally important and on the same footing, for a better humanity. In a certain sense, the Scimat Program can be seen as *Enlightenment 2*, but smarter. In this chapter, the motivation, significance and progress of The Scimat Program are outlined.

1.2 Prehistory

The story began with Thales (c. 624-c. 546 BC) who proposed the first "theory of everything": Everything is made of water. Subsequently and more prominently, Aristotle (384-322 BC) studied all aspects of the universe on equal footing, including astronomy, physics, biology, logic, ethics and politics. That is, he was interested in all the subjects now covered in various departments in the modern universities. This was not by accident at all.

There is a material basis underlying the fundamental unity of knowledge. Knowledge about our world/universe can be divided into two categories: those on nonhuman systems and those on humans. The former is what is called "natural science"; the latter, the humanities and social science. But this is misleading. Humans are *Homo sapiens*—a material system made up of atoms, just like those studied in "natural science". And since science is to understand Nature which includes all material systems, the inescapable conclusion is that humanities and social science are or should be part of science. In short, *everything in Nature is part of science* [Lam, 2008a]. If this point is not yet clear to everybody, it is due to the misconception or misuse of the word "science" in our daily language [Lam, 2014].

In fact, in the last 400 years or so since Galileo (1564-1642), modern "science" [1] has progressed rapidly because of three factors: scientists pick the simple systems to study; they make a lot of simplifications; they use external detectors and information processors (e.g., computers). Partly due to the great successes of these studies, these days for many people, the word "science" is implicitly and incorrectly identified with the "science of simple systems", while the "science of complex systems" to which all human-dependent knowledge belongs is often neglected.

1.3 Scimat

Scimat (Science Matters) is a new term coined by Lui Lam in 2007/2008 [Lam, 2008a]. *Conceptually*, Scimat represents the four tenets that (1) Science is humans' effort to understand Nature without bringing in God or any supernatural; (2) Science covers everything in Nature; (3) Nature includes humans and all nonhuman systems; (4) all research on human matters, humanities in particular, are part of science. (If the humanities do not look like science to many people, it is because they are mostly still done at the empirical level, one of the three research levels in any discipline—empirical, phenomenological and bottom-up levels.)

Disciplinarily, Scimat represents the collection of research disciplines that deal with humans; thus, as a branch of science,

Scimat = Humanities + Social Science + Medical Science

and

Science = Scimat + "Natural Science"

where "Natural Science" means the science of nonhuman systems.

Scimat's tenets result from two recognitions: (1) Humans are a material system (made up of atoms); (2) Science is the study of Nature which includes all material systems. The first recognition follows from

---

[1] In this chapter "science" with double-quotation marks means science in the conventional sense, which is the sum of "natural science" and social science but excludes the humanities (see Lam [2014]).

Darwin's evolutionary theory (1859) and Einstein's work on Brownian motion (1905). (For more see Lam [2014].)

The viability of scimat has been demonstrated in these four areas: history, arts, philosophy, and science studies. In particular,

- Human history is shown to be a science, too (confirming the historian Robin Collingwood (1889-1943) [1922]). Different scientific approaches/techniques to do history, apart from the usual narrative approach, are pointed out with examples. A historical law is discovered in the lifetime of Chinese dynasties [Lam, 2008b].
- The origin and nature of arts—a 2,400-years-old puzzle since Plato—is solved [Lam, 2011].
- "Philosophy", the discipline in humanities today, is shown to be a science matter [Lam, 2016].
- The confusion and misconceptions about Science is clarified by a historical examination of its development. A historically correct definition of science is offered, viz., Science is humans' pursuit of knowledge about all things in Nature *without* bringing in God or any supernatural, guarded by the reality check; Science thus has two components: scientific process and scientific results [Lam, 2014].

1.4 The Scimat Program

The Aristotelian tradition of treating everything in the universe as equally important and from the same footing was broken after the Renaissance. The rapid and tremendous success of modern science, largely in the "natural sciences", led to the specializations of the professionals. The compartmentalization of knowledge into different disciplines occurred just a few centuries ago [Turner, 2014], more by convenience than by the intrinsic nature of the knowledge involved.

Concerted efforts to revive this tradition all failed with one exception: the *Enlightenment* (1688-1789) [Porter, 2001]. The Enlightenment succeeded in ushering in social science (by establishing

Economics through the effort of Adam Smith) but failed in turning the humanities into a science, due to its insufficient understanding of the human system which is an open, complex system. It thus succeeded only partially with its aim of creating a "science of man", or, a Human Science. It was a movement interrupted and unfinished.

Individually, the book *Consilience* by Edward O. Wilson [1998], a Harvard biologist, advocates the revival of the Enlightenment spirit, based on other considerations. Apparently, it has failed to turn the tide, due partly to the lack of an action plan.

The Scimat Program, started by Maria Burguete and Lui Lam in 2007, is the latest international effort to revive the Aristotelian tradition of a unified knowledge and a continuation of the Enlightenment movement, working towards a Human Science. It is an *action plan* and is the "only game in town". Its emphasis is on raising the scientific level of the humanities (by encouraging the collaboration between humanists and "natural scientists") because decision making, a branch of humanities, is what determines the well-being of the human race. Thus, the ultimate aim of the Scimat Program is for a better humanity (Fig. 1.1). The Scimat Program consists of six steps.

<u>Step 1: A biennial international scimat conference series was set up</u>. Four conferences have been held so far, all in Portugal; the first three cochaired by Burguete and Lam; the fourth by Burguete and Nigel Sanitt; the fifth one to be held in 2015 (Fig. 1.2).

Fig. 1.1. The ultimate aim of the Scimat Program: Let the Earth be peaceful forever!

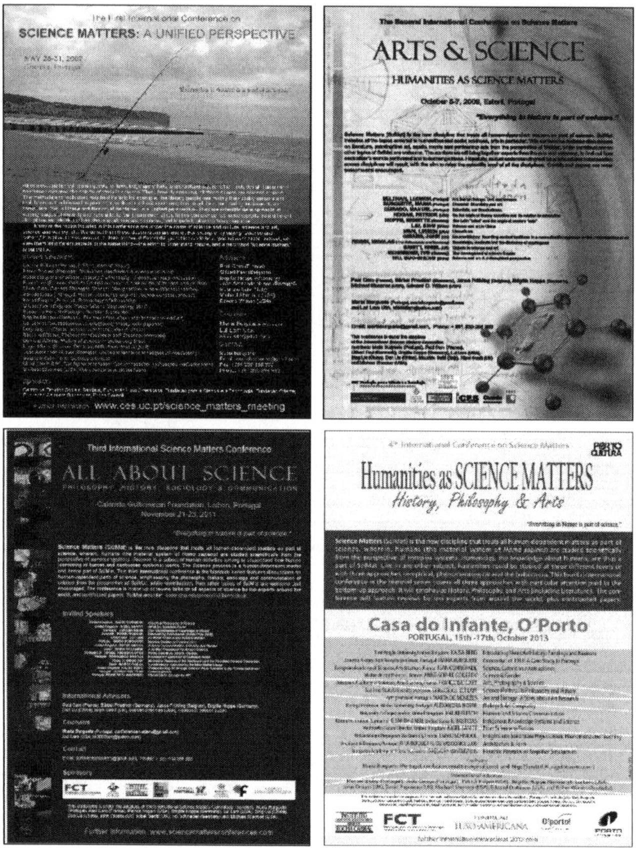

Fig. 1.2. Posters of the first four international scimat conferences, held all in Portugal, in the years, respectively, 2007, 2009, 2011 and 2013 (from left to right, top to bottom).

<u>Step 2: International Science Matters Committee (ISMC) was set up</u>. The ISMC was established in 2007 to promote the scimat idea and oversee the international Scimat Program (Fig. 1.3).

<u>Step 3: The Science Matters book series was set up</u>. Publisher: World Scientific; Founder and Editor: Lui Lam; three books are published (Fig. 1.4).

<u>Step 4: Scimat Centres being established</u>. We are in the process of establishing a number of scimat centres (100 eventually) around the

world (which would be independent from but collaborating with and reinforcing each other). The Centre is:

1. To do fundraising to support the Centre financially.
2. To organize international workshops/conferences and summer/winter schools.

1. Manuel **Bicho** (Portugal)
2. Peter **Broks** (UK)
3. Maria **Burguete** (Portugal)
4. João **Caraça** (Portugal)
5. Paul **Caro** (France)
6. Jean-Patrick **Connerade** (UK/France)
7. Patrick **Hogan** (USA)
8. Brigitte **Hoppe** (Germany)
9. Lui **Lam** (USA), *Coordinator*
10. Bing **Liu** (China)
11. Dun **Liu** (China)
12. John **Onians** (UK) — Father of Neuroarthistory
13. David **Papineau** (UK) — President, British Society for Philosophy of Science (1993-1995)
14. Kok Khoo **Phua** (Singapore)
15. Nigel **Sanitt** (UK)
16. Ivo **Schneider** (Germany)
17. Michael **Shermer** (USA)
18. Robin **Warren** (Australia)

President, International Union of History and Philosophy of Science (2009-2013)

President, European Academy of Sciences Arts & Letters

Author, *Cognitive Science, Literature and the Arts*

Columnist, *Scientific American*; Editor, *Skeptic* Magazine

Nobel Laureate (2005)

Fig. 1.3. Members of the International Science Matters Committee (ISMC). Presently, there are 18 members coming from five countries, which include a Nobelist, president of a European Academy, research institute directors, prominent magazine columnist and publisher, and eminent professors who are pioneers in their research fields. See Appendix 1.1 for their biographies.

Fig. 1.4. The Science Matters book series and the three scimat books published (in 2008, 2011, 2014). The first book *Science Matters* has been translated and published in Chinese (2013) and Portuguese (2015).

3. To communicate the scimat ideas to the public.
4. To give out an Award every two years, in the donor's name perhaps, for an individual who contributes significantly in the advancement of scimat.
5. To host short-term visiting scholars who will give lectures/short courses, who will also collaborate with existing faculty members and students of any discipline, especially from the humanities.
6. To help match faculty members from humanities and science departments, and give them release time to create new interdisciplinary courses (e.g., Science of History).
7. To help promote the new general-education course on "humanities + science" (Fig. 1.5) for undergrads of all majors.

Note that the Centre will not do research within itself, and so the maintenance fee is very minimal. With enough (outside) money, it can advance scimat by funding interdisciplinary research within a university. The scimat centre will be in a leading position academically in the most important multidiscipline of the 21st century.

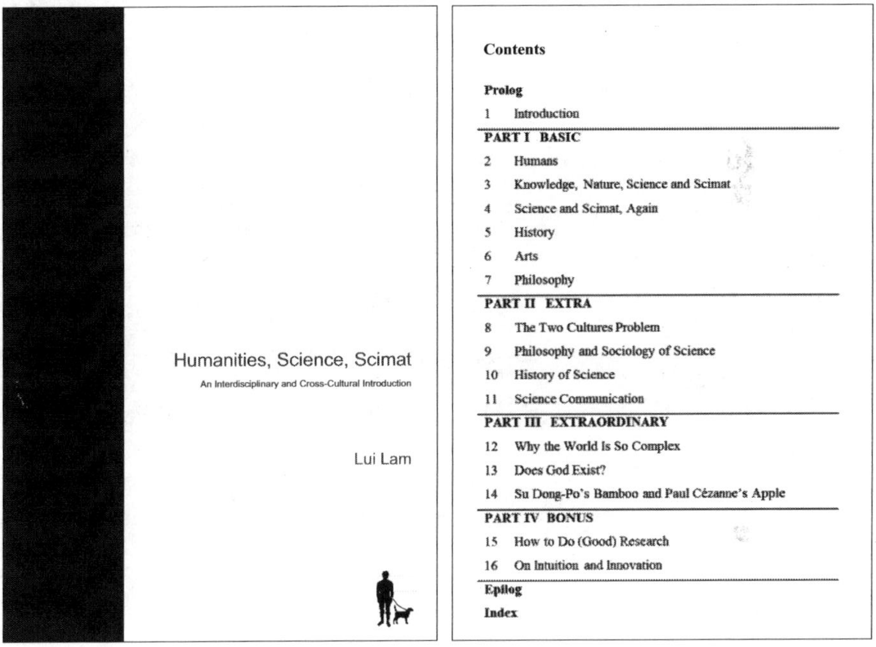

Contents

Fig. 1.5. The new general-education course for all majors and everybody else [Lam, 2017].

<u>Step 5: Establish an international scimat society</u>.

<u>Step 6: Publish an international scimat journal</u>.

We are now working on step 4; steps 5 and 6 are for the future, hopefully the near future. For more see website: www.sjsu.edu/people/lui.lam/scimat.

1.5 Conclusion

Scimat's motto is "Everything in Nature is part of science". Scimat's key insight is that we have "*One culture, two systems, three levels*"—science culture, simple and complex systems, three research levels [Lam, 2017]. What we are witnessing here is the initial stage in the birth of a new "multidiscipline"—more precisely, a new paradigm—called Scimat. It is similar to the case of History of Science (initiated by George Sarton early last century) and of Artificial Life (by Christopher Langton in 1986), but not quite. Scimat is much larger in scope since it incorporates the research of everything related to humans, and thus will be more far reaching in its influences. In particular,

- It provides a unified perspective for all the disciplines in the humanities, social science and "natural science".
- It is a rally point to raise the *scientificity* (i.e., scientific level) of the humanities, making the world a better place (since many large-scale human tragedies can be traced to the underdevelopment of the humanities in the last 2,400 years since Plato).
- It is the foundation behind the "synthesis" of the humanities and "science", solving the so-called two-culture problem at the basic level.
- It provides the basic rationale for general education and a route to make it successful.
- It provides the broadest framework in interdisciplinary learning/ teaching, and science teaching [Matthews, 2015].
- It is the most interesting and important multidiscipline in the 21st century.

In short, Scimat advocates the understanding of our world through science and rational thinking, whereas humanities are recognized as part of science. Let us work together for a better humanity and make the world a peaceful place forever, for us and our children!

Appendix 1.1: The International Science Matters Committee

On May 30, 2007, in Ericeira, Portugal at the end of the First International Conference on Science Matters, an International Science Matters Committee (ISMC) was formed; it consisted of 9 members. Since then the ISMC was expanded. After the first four scimat conferences in the biennial series, at this point, there are 18 members in the ISMC (see Fig. 1.3). Here are their brief biographies.

Manuel Bicho obtained his PhD in Medicine (Physiology and Biochemistry) in 1985 and MD in 1975, both from Lisbon Medical School, University of Lisbon. He is Director of Rocha Cabral Institute, and Director of Genetics Laboratory (since 1995) and Professor of Genetics, Faculty of Medicine of Lisbon (since 1990). His main research area is Biochemical and Molecular Genetics. He was awarded the Ernesto Roma Prize in Diabetes Research (1996) and the Bial Prize of Clinical Medicine (1994).

Peter Broks obtained his BA and PhD from the University of Lancaster. In 1990 he was a specialist appointment at the University of the West of England, Bristol, to design, develop and teach a new undergraduate programme in "Science, Society and the Media" jointly run by the Faculty of Humanities and the Faculty of Applied Sciences. He has published extensively in the history of science especially as it relates to popular culture and is the author of *Understanding Popular Science* (2006). In August 2011 he left UWE so that he could devote more time to research and writing.

Maria Burguete got her PhD in History of Science (contemporary chemistry) from Ludwig Maximilians University at Munich, Germany (2000), after graduation in Chemistry & Biochemistry at Faculty of Sciences, Lisbon University and a BA in Chemical Engineering. She is a scientist and a university lecturer with teaching and research experience in a wide variety of scientific fields. She is a scientific researcher at Scientific Research Institute Bento da Rocha Cabral in Portugal since 2007. She has published 11 books in the scientific field (and seven books in poetry) and over 25 scientific papers. She is a corresponding member of European Academy of Sciences, Arts & Letters since 2010.

João Caraça obtained the D. Phil. in Nuclear Physics at the University of Oxford (1973). He is the Director of Calouste Gulbenkian Foundation's Delegation in France, and Full Professor of Science and Technology Policy at the Instituto Superior de Economia e Gestão of the Universidade Técnica de Lisboa. He is member of the Governing Board of the European Institute of Innovation and Technology (EIT). He also integrates the Steering Group of the European Forum on Philanthropy and Research Funding and is President of the Advisory Board of the Portuguese Business Association for Innovation (COTEC). Caraça was Science Adviser of the President of the Portuguese Republic (1996-2006) and has published over 150 scientific papers. His main interests are science and technology policy and prospective studies. His recent books include *Limits to Competition* (1995), *Science et Communication* (1999), *Entre a Ciência e a Consciência* (2002), *Le Printemps du Politique* (2007), and *Ideias Perigosas para Portugal* (2010).

Paul Caro is a former (retired) Director of Research at CNRS who has worked for many years in inorganic chemistry. He is a rare earths specialist. In the 1980s he became interested in science popularization through newspaper articles (in "Le Monde" and magazines), radio broadcasts (France Culture, Radio Classique, mostly), television shows (TF1), exhibitions in Museums and some books. He was until 2001 in charge of "scientific affairs" at the Cité des Sciences et de l'Industrie in Paris. He is a Corresponding Member of the French Academy of Sciences and a Member of the French Academy of Technology.

Jean-Patrick Connerade Educated Lycée Charles de Gaulle and Imperial College D.Sc. University of London 1976. Lockyer Professor of Physics for many years, now Emeritus Professor and Distinguished Research Fellow Imperial College London, Hon Prof Physics East China University Shanghai and Permanent Visiting Prof. WIPM Chinese Academy of Sciences. President of the European Academy of Sciences Arts and Letters, Honorary President of Euroscience. Some 250 published papers. He is also a published poet in the French language, recipient of several literary prizes (prix Jose-Maria de Heredia de l'Académie Française, prix Paul Verlaine de la Maison de Poésie de Paris, Grand Prix de Poésie de la Société des Poètes Français).

Patrick Colm Hogan received his BA in Philosophy from Santa Clara University, his MA in Philosophy from the University of Chicago, and his PhD in English from the State University of New York at Buffalo. He is the author of 14 books, including *The Mind and Its Stories: Narrative Universals and Human Emotion*; *Cognitive Science, Literature, and the Arts: A Guide for Humanists*; and *Understanding Indian Movies: Culture, Cognition, and Cinematic Imagination*. He recently edited *The Cambridge Encyclopedia of the Language Sciences*.

Brigitte Hoppe obtained her state diploma in pharmaceutical and biological sciences from the University of Freiburg im Breisgau, her degree of Dr. phil. nat. from the University of Frankfurt am Main (1964), finished the habilitation in History of Science at the University of Munich (1972) and became Associate Professor of the Ludwig-Maximilians-University of Munich in 1980, where she developed a working group in the history of life sciences and chemistry. She published 7 books and more than 200 papers on the history of sciences. She was a member of the editorial board of *History and Philosophy of Life Sciences* for 10 years and is now a member of the editorial board of *Archives Internationales d'Histoire des Sciences*. She is an Effective Member of the *International Academy for History of Science* and member of several national and international Societies for History of Sciences.

Lui Lam, humanist and physicist, obtained his BS (First Class Honors) from University of Hong Kong; MS., University of British Columbia; PhD, Columbia University. He did his PhD thesis at Bell Labs. Lam invented Bowlic liquid crystals (1982), Active Walks (1992), and two new disciplines: Histophysics (2002) and Scimat (Science Matters) (2007/2008). He has published 16 books and over 180 scientific papers; the books include *Introduction to Nonlinear Physics* (1997), *Arts: A*

Science Matter (2011) and *All About Science* (2014). He is the founder of the International Liquid Crystal Society (1990); cofounder of the Chinese Liquid Crystal Society (1980); founder and editor of two book series, "Science Matters" (World Scientific) and "Partially Ordered Systems" (Springer). Lam is an editor of *Physics* and *Science Popularization*. His current research is in scimat and complex systems; scimat website: www.sjsu.edu/ people/lui.lam/scimat.

Bing Liu obtained his BS from Peking University (physics department) and M.Sc. from the Graduate School of Chinese Academy of Sciences. Now he is a professor of history of science at Tsinghua University, vice director of the Center for Science Communication and Popularization of CAST and Tsinghua University, and a Guest Professor at some universities in China, such as Shanghai Jiaotong University. His current research include history of physics, historiography of science, philosophy of science, and science communication. He published 17 books (and translated 7 books, edited more than 30 books), over 130 academic papers, and many notes.

Dun Liu, former Director of the Institute for the History of Natural Science, Chinese Academy of Sciences (1997-2005) and past President of the International Union of History and Philosophy of Science (2009-2013), is currently Professor Emeritus of the Institute and Professor of Tsinghua University at Beijing. His main research field is Chinese mathematics/astronomy and its interaction with the social context, especially in the Ming-Qing transitional period (c. 17th century). Also serving as editor-in-chief of the bimonthly journal, *Science & Culture Review*, he currently focuses on such historiographic and cultural topics as the "Needham question" and the "C. P. Snow thesis".

John Onians, studied Classics at Cambridge University and Art History at the Courtauld and Warburg Institutes, London University, before being appointed Lecturer, Senior Lecturer and Professor at the University of East Anglia. He was founding editor of the journal Art History and founding Director of Research and Academic Programs at the Clark Art Institute, Williamstown, Mass and has been a Getty Scholar and held

Fellowships at the Center for the Advanced Study of the Visual Arts, Washington, and the Wissenschaftskolleg, Berlin. He has published numerous books including, *Art and thought in the Hellenistic Age: The Greek World View 350-50BC* (1979), *Bearers of Meaning: The Classical Orders in Antiquity, the Middle Ages and the Renaissance* (1988) and *Classical Art and the Cultures of Greece and Rome* (1999), and he edited the first *Atlas of World Art* (2004). He is now writing *European Art: A Neuroarthistory.*

David Papineau, professor of Philosophy of Science in the Department of Philosophy at King's College London, has a B.Sc. in Mathematics from the University of KwaZulu-Natal, and a B.A. and a Ph.D. in Philosophy from Cambridge University. He was President of the British Society for the Philosophy of Science (1993-1995) and President of the Mind Association for 2010. He was editor of the British Journal for the Philosophy of Science (1993-1998), a Leverhulme Research Fellow (1999-2000), and a Mind Fellow (2007). His books include *Theory and Meaning* (1980), *Philosophical Naturalism* (1993), and *Thinking about Consciousness* (2002).

Kok Khoo Phua was born in 1942 in Singapore. He obtained his BSc DIC from Imperial College, London and PhD in Mathematical Physics from the University of Birmingham, UK in 1970. He is a Fellow of American Physical Society. He is currently the Chairman and Editor-in-Chief of World Scientific Publishing Company. He is the Founding Director, Institute of Advanced Studies (IAS), Nanyang Technological University (NTU) Singapore and Adjunct Professor, Department of Physics, National University of Singapore (NUS).

Nigel Sanitt obtained his BS from Imperial College London and Part III of the Mathematics Tripos and PhD from Cambridge University. He trained as an astrophysicist at the Institute of Astronomy, Cambridge, before becoming a lecturer at University College, London University. He is founder and editor of *The Pantaneto Forum.* The journal aims to promote debate on how scientists communicate, with particular emphasis on how such communication and research skills can be improved through a better philosophical understanding of science.

Ivo Schneider is Professor emeritus for the history of science of the Universität der Bundeswehr München. He was professor and visiting professor at the universities of Munich, Princeton, Bielefeld, the university of Minnesota in Minneapolis and the technical university of Budapest. He is cofounder, editor, coeditor and reviewer of several journals and series of the history of science and the history of mathematics. Special research interests concern the history of classical probability theory, mathematical practitioners and reckoningmasters in the 16th and 17th centuries, scientific instruments in the 17th and 18th centuries, biographies of scientists, science theatre and the origins of Bavarian optical industry in the first half of the 19th century. His books comprise a source book of the history of probability theory up to Kolmogorov and biographies of Archimedes, Johannes Faulhaber and Isaac Newton. He is a member of different national and international societies for the history of science and ever since 1995 membre effectif of the Académie Internationale d'Histoire des Sciences. His autobiography was published in 2004 in a volume devoted to his 65th birthday. In the same year he was honoured with the degree of a Doctor honoris causa by the technical university of Budapest.

Michael Shermer is the founder and editor of *Skeptic* magazine, a contributing editor and monthly columnist for *Scientific American*, and the host of the Skeptics Distinguished Lecture Series at Caltech. He is the author of *Science Friction*, *The Science of Good and Evil*, *Why People Believe Weird Things*, *How We Believe, In Darwin's Shadow*, *The Borderlands of Science*, and *Denying History*. Shermer received his PhD in the history of science from Claremont Graduate School.

Robin Warren was born in 1937, in Adelaide, South Australia. He graduated MB, B. from the University of Adelaide in 1961. After training at the Royal Melbourne Hospital, he was admitted to the Royal College of Pathologists of Australasia in 1967. Since then, he was a senior consultant pathologist at the Royal Perth Hospital in Western

Australia, becoming emeritus consultant pathologist in 1998. In 2005 he (together with Barry Marshall) was awarded the Nobel Prize in Physiology or Medicine for the "discovery of the bacterium *Helicobacter pylori* and its role in gastritis and peptic ulcer disease."

References

Collingwood, R. G. [1922] "Are history and science different kinds of knowledge", *Mind* XXXI, 443-451.

Lam, L. [2008a] "Science Matters: A unified perspective", in *Science Matters: Humanities as Complex Systems*, eds. Burguete, M. & Lam, L. (World Scientific, Singapore) pp. 1-38.

Lam, L. [2008b] "Human history: A Science Matter", in *Science Matters: Humanities as Complex Systems*, eds. Burguete, M. & Lam, L. (World Scientific, Singapore) pp. 234-254.

Lam, L. [2011] "Arts: A Science Matter", in *Arts: A Science Matter*, eds. Burguete, M. & Lam, L. (World Scientific, Singapore) pp. 1-32.

Lam, L. [2014] "About science 1: Basics—knowledge, Nature, science and scimat", in *All About Science: Philosophy, History, Sociology & Communication*, eds. Burguete, M. & Lam, L. (World Scientific, Singapore) pp. 1-49.

Lam, L. [2016] "Philosophy, Science and Scimat", in *Humanities as Science Matters: History, Philosophy and Arts*, eds. Burguete, M. & Riesch, H. (Pantaneto Press, Luton, UK).

Lam, L. [2017] *Humanities, Science, Scimat* (World Scientific, Singapore).

Matthews, M. R. [2015] *Science Teaching: The Contribution of History and Philosophy of Science*, 20th Anniversary Revised and Enlarged Edition (Routledge, New York).

Porter, R. [2001] *The Enlightenment* (Palgrave, New York).

Turner, J. [2014] *Philology: The Forgotten Origins of the Modern Humanities* (Princeton University Press, Princeton).

Wilson, E. O. [1998] *Consilience: The Unity of Knowledge* (Alfred A. Knoff, New York).

2

Combining Permanence and Change in Higher Education: The Strength of Strong Values

Filipa M. Ribeiro

What do Mark Edmundson, English Professor at the University of Virginia (USA) and Daniel Pennac, a French teacher and writer, have in common? In different ways they address what is at stake in any educational system or level: individual and, then, structural transformation. Currently, universities are lost in the translation of labels such as "elite universities", "corporate universities", "entrepreneurial universities", "academic capitalism". Nevertheless, teaching and learning are still the backbone of any endeavour to resist the confluence of forces in higher education that are leading to greater conformity and consumerism at the expense of inquiry, inspiration and challenge. As any transformation process, teaching and learning must have a permanently grounded basis to orient and guide curriculum and personal development within a community. This paper proposes that this common ground should be based on a theoretical orientation drawn upon the idea of sustainability and preparation of the future, with a significant emphasis on the human rights, on the ethics of the Earth, on development and culture. Thus, this paper aims at: (1) presenting and explaining each one of these ideals considered as structural social factors; (2) justifying these values with empirical evidence from 30 knowledge networks from scholars based in Catalonia. The purpose is to show how strong teaching and learning in any field must share some permanent and transdisciplinary values and ideals so that universities become places of a true ubiquitous education where everything that is taught and learnt cracks the shell of convention.

2.1 Introduction

The division of organized knowledge hinders a fuller understanding of the world and the evolving needs of the individual. The task before us is to rebuild our institutions so that we can maximize opportunities to further self-knowledge and, by extension, to widen a context-rich collective knowledge. This is a huge task more because of the conservatism inherent within higher education institutions than because of the lack of willing and capable people to pursue that task. But the recognition of the enormity of the task does not mean that it is impracticable. We are currently facing a "multifurcation" of the structures of knowledge that appears to be chaotic in many ways, from which we ought to emerge knowing that the order is not determined, but determinable [Wallerstein, 2003].

The well-known "science wars" focused mainly on the nature and validity of knowledge that produces and legitimises the changing world through science. It was essentially a debate among scientists [Santos, 2003]. It is, thus, necessary to broaden that debate to other arenas, such as Higher Education (HE), namely in one dimension that is continually neglected: values and ideals. Are institutions capable of making the choices and acting in such a way that we can have a "sustainable", coherent and non-exclusive knowledge? Or are we doomed? Will there be a Darwinian solution? What lessons from our past may provide insights on our capacity to make the kind of sacrifices necessary? What are the limits of institutional action in the promotion of values and in the promotion of the educative value of knowledge?

The public responsibility of higher education and research is a cornerstone of the European university heritage. Yet, our societies is changing rapidly and clinging to old solutions based only on the fragmentation of knowledge and experts to a productivity age are not enough. However, evoking any public responsibility of higher education institutions forces us to look more closely at different degrees and levels of public responsibility as well as the instruments available for exercising such responsibility. This (moral) relativism urges us to ask: what can we rely upon? Answering this question is even more complicated when we start thinking outside our "social box". Thinking globally suggests that there could be multiple valid moral codes or belief systems, yet we wouldn't want to say that people around the world have different intuitions. So, we are back to moral relativism, which, in Higher Education, enhances and is enhanced by policy and political relativism. Bain-Selbo maintains that:

> "(…) democracy is not a cold, legalistic framework in which our religious/moral communities exist – all separated from one another with nothing in common. Democracy is our tradition. It is our moral community – shaped by legal principles, but also shaped by a long and rich history. It is that place where we can practice moral virtues and develop moral character. It is the foundation upon which we can stand to speak the truth" [Selbo, 2008, p.12].

We argue that this foundation, though valid, has proved to be incomplete mostly because it neglected the basic starting point of any educational endeavour: the "acquisition of knowledge, along with the development of rationality, is still important, in fact, the primary justification for educational experiences" [Drewe, 2001, p.235]. Thus, higher education institutions (HEIs) need to know 'what they know' and how and what they are using its knowledge for. The emphasis on organizational culture has underscored the integrative role of management [Henkel, 1991] and on the extremes of the individualization of HE actors [Stoer & Magalhães, 2003], the attempt to conciliate a broader set of values in HEIs and in HE policies and governance must deal with the fact that the new dimensions of students are not confined to their individual skills, but with oriented ideals and goals.

2.2 Ideals as social facts

Until now sociology has closed the circle of the "individual society", but now it is being called upon to extend the scope of their research and broaden its conception of the "social fact". This Durkheimian object reveals its hidden face expanding its size when we linked it to its aspirational potential. René Hubert offers us the neo-Kantian thesis of critical relativism on the development of a critical consciousness. He supports his theory on a part of an essay by Octave Hamelin about the essential elements of representation. I will not consider the full argument of this thesis, but I will try to apply to solve the problem posed by this paper: the strength of strong ideals and values to higher education.

Hamelin gave a new meaning to the dialectic by erasing the aggressive aspect of the struggle between opposites. Hubert transforms it into an instrument of pedagogical action, in order to explain and guide the consciousness's development. This development is the same process of education in a world where our experience clearly shows us how life gives us the mastery of all external reality through the evolution of the spirit. I will briefly explain how this process occurs. Firstly, we have the theoretical consciousness that is born from our relationship with the world [Damásio, 2012]. We are the subject and the world is the object. We grasp and internalize the external reality by formulating categories of reason. These categories are our own experience of things. The experience gives us the worldview or view of the world. But the subject-object relationship is transformed into our consciousness in recreating the world in ourselves and therefore in the recreation of our own consciousness, which is renewed in relation to the world. The contradiction between subject-object disappears due to the fusion of subject and object in consciousness. Therefore, the representation of the world in us is not simple overlay of external reality, but an absorption and assimilation of the real. Knowledge ceases to be information and data collection and becomes a lived experience. The theory of consciousness – in relation to the world – reshapes its own position before the world and identifies itself with reality. This initial identification produces what Adolf Ferriere [1955] named as "remelting of personality". The personality is modified at the same time that our subjective and objective view of our own capacity for action in the world is changed. This actually helps us to understand the thesis of learning by doing, defended by Renouvier [1881] in the *Le sentiment de l'effort.* At this point we move to the order of phenomena where we handle the things and beings through science and technology. Science gives us the theoretical consciousness; the practical one leads to the consciousness's activity, no longer as a simple experience, but as a recreation. We recreate the world while we recreate ourselves. That's when we reach the aesthetic consciousness, the final synthesis of the dialectic of consciousness. The aesthetic capture of the world becomes a representation of the total reality and our inner reality (as recent fields show, such as user-experience in design of applications and mimetic technology). The aesthesy is then defined by our own desire to become something else. This feeling has led us to reason and is behind proposals such as the one of Andreas Reckwitz [2002] who deals with the problem of where to place the material in relation to the symbolic. According to the German sociologist, that place shouldn't be neither "culturalist" nor "materialist". One of the developments of this trend is the recent cultural theories described as "Theories of social practices". To Reckwitz, the practice theories ascertain that social order is formed in the crucible of cognitive-symbolic relations. We argue that ideals are a step forward in relation to these relations. Fawns [2008] argues that high-modern culturalism defines materials as "objects of knowledge" or "symbolic objects" as objects that become visible and, in the field of education, that vision on materiality has been the dominant one. Ideals as social facts contradict this dominance as ideals for their own sake sought to place education beyond value, beyond the messiness of the market and everyday life. But in our age no activity is beyond the reach of value and we must ask again what is the real and irreducible value of the ideals in our lives and culture, and make sure we play our part in the wellbeing of society as a whole. Ideals are something real and efficient, but discrete and they do not oblige individuals; rather they suggest their own importance giving us the freedom to accept them or reject them. However, it is important to say that preaching an ideal has no value. But each ideal is to be assimilated so that it does not get compromised under pressure of greed, pride, infatuation and so on. So, deliberate living of a chosen ideal is imperative to complete the process of assimilation.

Ideal, just as value, indicates the regard for a thing, situation, or attitude, which for some reason is esteemed or prized by the ideal holder and to others. In this paper, ideals refer to certain universal current ethical attitudes. It is important to notice that ideals are not just arbitrary man-made rules but stem from an inherent, common regard for one's own interest. There may be some cultural variations in degree or emphasis but the basic standards have certain universality.

2.3 Why bring back ideals into HE?

Why are ideals as social facts needed? Firstly, to overcome the effects of the atrophy of ends and of the hypertrophy of means and to better tackle the educational imbalances around the globe. Secondly, because development has to be diverse with *ad hoc* strategies oriented towards a defined end. Ideals in HE can actually be rewarding if applicable into the criteria to evaluate scientific achievement. Thirdly because no one living in this world can escape relationships and failure to meet ideals and values puts one in conflict with oneself.

In the past, HEIs were expressions of well-assimilated and dominant social values, but those ideals were mostly institutional values and not so much individually assimilated values. For that to happen, it is necessary that researchers assimilate personally and professionally the ideals that lead to a common good. In other words, bringing back ideals to HE is not only about policy change but it is about behavioural change – by researchers working with national and local authorities, with business and educationalists to recognise and increase how educational and scientific contribution can enrich society and enhance our cultural identity. How can rhetorical commitments to new forms of leadership, innovative practice and generous collaboration turn into something real? This is where ideals and researchers can come in. Their ethos, their method, their creativity can act as the catalyst for new ways of being and thinking.

2.3.1 Which ideals?

Given that this chapter argues that ideals are ethical imperatives that should bring together the efforts both of natural sciences and humanities, we now present for discussion a list of such ideals. These ideals stem from a theoretical orientation within a symbiotic evolution placed between the idea of sustainability and the preparation of the future, with a significant emphasis on human rights, on the ethics of the Earth, on development and culture based on the joint contribution between science and humanities.

2.3.1.1 Biodiversity

Biodiversity is the conjugation of fraternity and equality extended to all beings on the planet. "Politics and the economy are advanced symbioses in which the original competition of all against all has been replaced, in part, by cooperative mechanisms with ethical contents", says Aldo Leopold in his Ethics of the Earth. That should continue to be so, or more properly, to start being so since we have strongly despised both fraternity and equality and we omit biodiversity as an intrinsic value to all civilized human activity. Above all, without biodiversity, freedom is just an irony and the most effective way to subdue the people, forcing them to mortgage the future on someone's behalf. We have examples of unavoidable consequences of that mortgage, this iceberg whose visible part is called financial crisis.

2.3.1.2 Societies are all of us

Family, work and equanimity. Availability, objectivity and clarity. Initiative, continuity and evaluation. Citizenship, responsibility and empowerment. Love, well-being and freedom. Affection, motivation and knowledge. Humanity, wisdom and progress. Hypothesis, verification and completion. Here are the different outlets (of consciousness, stadiums or values) which have always attended the demand of culture and civilization on the barbarism of the abstract, of mediocrity, of superstition, treacherous and dishonest conservatism's, of the malicious ignorance, and fundamental stiffness. Everything can be recycled, including the essential foundations of our identity and culture. To recycle our inner child in us, as individuals and as a society to get to the next stage of evolution and progress is not a utopia, ideology or religion: it is an attitude. It is a new attitude towards life, which comes from observation, reflection and practice of many groups and people around the world who realized that we are going through a serious civilizational crisis. To solve this crisis it is necessary to demolish old (pre) concepts, values and powers, as well as to define alternative paths that cherish the individual and collective creativity, that respect life, the environment, the nature, the ecosystem, the atmosphere, the water and the human condition.

2.3.1.3 Evolution is action

The difference between acting and reacting is notorious. Acting is participating; reacting is to submit to. One is no longer an aware subject when he just reacts because, in that circumstance, he is just responding to who has manipulated him. Only the one who acts is truly (r)evolutionary. Through scientific progress and human development, we gained a new understanding of the world and, through the sensorial sense, we imported from nature core values and non-human technological or behavioural models. The civilisation behaviours socially advanced, like how to recycle urban solid waste, how to increase net primary productivity, how not to produce genetically modified organisms, how not to introduce chemicals into the environment, how not to feed wild animals in addition to the domestic, how to protect and promote conservation and restoration operations of plots of land and ecosystems.

There is also the need to include the demarcation of natural parks and reserves, to restructure the curriculum body and university educational and programmatic, to reform the Administration and technological planning, aiming at a better human physical occupation of the territory.

2.3.1.4 Balance or eternity?

Until a few decades ago, and due to several ideologies, mankind understood the environment under an anthropocentric perspective in which nature is presented only as an instrumental value associated with a range of potential and benefits, including aesthetic, historic, scientific and economic. Nature was a source of encouragement for creativity and imagination, as well as a significant contribution to the integral development of the human being. As a result of progress in the well-being and living conditions, in science and technology, mentalities have evolved, and today the overall trend lies in an understanding of the environment under two distinct, though simultaneous, perspectives: ecocentrism and biocentrism. With Copernicus and Galileo, the Earth is no longer the centre of the universe; with the Brundtland report [1987] the era of human supremacy on Earth ended as well. This shift had irreversible effects, and forced to recycle all the theories of understanding the planet. This understanding is now more lucid, objective, rigorous, logical, ethical and functional, but also more broadly in order to recognize that without the preservation of the diversity of species it is impossible a sustainable balance of the planet.

According to the biocentric perspective, we are just one of the strategies of life to become eternal. To the ecocentric perspective, we are simply members of a species that interacts with the others in an attempt to keep a balanced survival ecosystem. In practice, the two complement each other and require us to observe carefully our conduct and our technological deeds. The choice is individual, but each individual's choice will have (irreversible) consequences in the lives of others.

2.3.1.5 Biodiversity and the butterfly effect

> It is obvious that the economic viability limits of what may or may not be made in favour of the Earth. It has always been like that and always will be. The fallacy that the economic determinists tied around our neck and that we need now to eliminate is the belief that the economy determines ALL land use. This is simply not true. A lot of actions and

> attitudes, including maybe most of all relations with the Earth, are more determined by the tastes and predilections of users than by their wallet. Most of all relations with the earth depends more on investments of time, forecasting, capacity and faith than on investments in money. With regard to the use of the land, we are what we think [Leopoldo, 1949].

If there is an undeniable intrinsic value in each of the known life forms (biodiversity), regardless of the interest they may have for the human species, the non-exploitation of holistic entities like ecosystems and the Ecosphere, is the only plausible attitude to keep and recover global sustainability. Therefore, to strive for a direct, participatory democracy requires universal access to information and communication technologies, the collective awareness of the facts and problems of each and everyone, as well as a different accountability of citizens of their habitat.

2.3.1.6 Specism and multiculturalism

Although the human species has been the one who better got emancipated, every form of life is idiosyncratic, rare, unique and inimitable. The singularity must observe attentive care in consideration and compliance with the conditions of its flourishing, as well as the balance of the habitat that integrates it. In order to counteract the natural difficulties of expansion and survival Man has always practiced environmental manipulation through the specism, regarded as a sensible strategy. However, today we know that this strategy causes us troubles and irreparable damage, both to the ecosystem and the economy of the regions, both in the ecosphere as in the living conditions of each individual. The entrapment of ecosphere sustainability is much larger than the benefits received, since monocultures are more vulnerable to the effects of natural disasters and climate change. When a population is dependent on only one product (like technology, for instance), it is submitted to the least fragility of its reproduction.

It is time to reverse the course of decisions and reject any and all cultural practices which jeopardise the integrity of the fauna and flora of a place, or its territorial and geographical particularities. We must definitely opt for a widespread option of attitudes and behaviours that spread and disseminate the diversity (multiculturalism) and cultural specificities. The most rich fruitful and sustainable environments (and nations) are not those who boil down and close upon themselves, but the ones that change and multiply in accordance to the biodiversity and the largest possible number of opportunities, mobilizing all its endogenous resources to promote cooperation and autonomous development.

2.3.1.7 Ethics and instinct

The main concern of a responsible and critical citizenship is to ensure the right of everybody to a sustainable and balanced environment. Until very recently the human relationships with the earth, in addition to the usual patriotic and territorial quarrels, were strictly a synonym of many privileges and no obligations. Owners, tenants, business people, tourists, workers and residents, each in their own way, extensively instrumentalized the earth. The earth became an appendix of income and capital gains. It was forgotten that ethics is unambiguously a type of community instinct similar to the instinct of defence. The crisis that we are experiencing has compounded this relationship making considerable setbacks in various sectors placed between environment with economic interests and cultural traditions. This has already mirrored some improvement in quality, nature and intensity of the instrumentalization of the earth. Therefore, to rethink the use of arable soil, to establish priorities, the optimization of choice on the seeds, to hold and protect the seedlings of the locations prone to landslides, flash floods or fires, to promote agricultural and livestock research, to protect native species, as well as to redeploy the conduct of three r's (reduce, reuse and recycle), presents itself not only as an economic necessity, but also chaos as an ethical imperative.

2.3.1.8 Environmental policy and development

There are huge mistakes in environmentalism and democracy: the exploitation of man by man, the mitigation of the rights, freedoms and guarantees, the savage productivity and entrepreneurial sectors that only claim for more velocity and profit, the irrational exploitation of natural resources, the institutionalization of the dependencies, the conservation of perishable and inadequate short-term planning, especially when this is imposed by the subordination of political will to economic interests or ideological fundamentalisms. These errors have been continuously and consecutively repeated by the keepers of the State apparatus. There were in fact neglecting gaps since the middle of last century. These gaps occurred with the exquisite and subtleties of wickedness to skirt the guidelines of the Kyoto Protocol, which goals were the sustainability and the appeasement of more or less war conflicts and party-political divisions. In fact, to explore just to conserve or preserve just to explore, such as the establishment argues, is a fallacious and malicious panacea that aims solely at evading the responsibility, the social conscience and citizenship. In an ever-changing world, keeping anything unchangeable is just violating their nature. And we all know that, in environmental terms, since the 60s, you can only save what is improved or nothing at all.

2.3.1.9 Economy and society

Economical development is the direct "provider" of the satisfaction of desires and needs of a population or community; but it is not the only one. This satisfaction requires the human interference towards the integrity of the ecosystems. Paradoxically, those who are most dependent on biodiversity are also the most conducive ones to molest its integrity. The welfare of a species is almost always done with the sacrifice of the other, and mankind is not an exception to that verdict. However, sustainable development is one of the biggest steps taken by modernized societies in order to reconcile the exploitation of the environment without harming or jeopardizing the growth of the economy. It did so by aligning plans and budgets under the auspices of a consecutive optimization of results. The model of society which is based on a centralizer gigantism, whose socio-economic structures are highly complex and bureaucratic, prevents the observance of any type of sustainability, including the economic, as the current crisis shows. This is so because it facilitates the waste of energies and the concentration of powers, the hegemonic control of the processes of production and consumption, erasing the conscious participation and citizen responsibility on daily decision-making about their own life. Hopelessly, the individual is prevented from choosing for a harmonic and symbiotic relationship with the ecosystem in which he belongs to. Therefore, we must strive for a society that is both plural and polycentric; that is continuously built and rebuilt from the principles of diversification. Economy can promote society, and society should reinforce the ontological unity between the human being and the environment that surrounds it aiming at a perfect symbiosis. This symbiosis should be based on maximizing the opportunities for well-being, personal, social and professional realization of each individual. That's also the heuristics of sustainability: to provide ever more consolidated and effective sustainability - or sustainabilities.

2.3.1.10 Aesthetics and culture

When Aldo Leopoldo first launched his book *The ethics of the Earth* he claimed: “We abuse land because we see it as a commodity belonging to us. When we see land as a community to which we belong, then we can begin to use it with love and respect. There is no other way for the Earth to survive the impact of the mechanized man, and for which we can remove the aesthetic cropping with which culture united to science can contribute” [Leopoldo, 1948].

Ethics, culture and ecology (as a science among the panoply of many others) are not separate concepts in nowadays. They were not separate concepts already at the end of the first half of the last century. However, in that time they were merged by a sensible choice; instead today they are forced to be merged because we, as humankind, have no other alternative. If culture does not encourage a better understanding of us and of our surroundings, it will isolate us, makes us sick or pathological beings. Culture as a social reality is the synthesis of all human creativity, in an individual and collective dynamics. Culture aims at mobilizing and facilitating relations between people, between these and the region where they live, and the region they inhabit with other regions of the globe. Culture is not an expression of technocracy under the postulates of economisms that are incompatible with autonomy, with independence, with emancipation, with awareness and with decentralisation. Rather, it is the expression of diverse ways of living, of symbolic values and practices which have gained an aesthetic and affective importance for individuals and groups.

2.4 Researchers' ideals

Given the reflection on the purpose and place of ideals in education on the previous section of the paper, there is a need to know the actual relationship between ideals and the perceptions of scholars and scientists about their own work. 1684 researchers from all over the world were asked to answer a brief open questionnaire. The main purpose was to obtain information about the ideals of these scientists. The sample population is composed of researchers from the natural and social sciences affiliated to universities or other R&D institutions at PhD level. The data sample is composed by 388 scientists and it uses methods of stratified sampling by types of sciences (natural, social and human) and by geographical location (by continent). It also uses a random sampling method in the selection of scientists, incorporating the widest possible range of disciplines. The data about the participants' contact information and affiliation were obtained through three sources: 1) papers published in scientific journals from the Public Library of Science, namely Plos One, Plos Genetics, Plos Clinical Trials, Plos Medicine, Plos Computational Biology, Plos Biology, Plos Pathogens; 2) the websites of the universities randomly chosen; 3) a personal contact list. These sources were chosen because they were the most convenient ones according to the needs of reliable information and data availability. Within the group of the natural sciences, inquiries were sent to the following countries: South Africa, Germany, Argentina, Australia, Austria, Switzerland, Belgium, Brazil, Canada, China, North Korea, Costa Rica, Croatia, Denmark, Scotland, Spain, USA, Finland, France, Greece, Netherlands, Hungary, England, Iceland, Israel, Italy, Japan, Kuwait, Malaysia, Malawi, Morocco, Mexico, Nigeria, Norway, New Zealand, Poland, Portugal, Russia, Singapore, Sweden, Taiwan, Uganda, Venezuela, and Zambia. Within the group of the social sciences, inquiries were sent to the following countries: Argentina, Australia, Brazil, Canada, Scotland, USA, Finland, France, Netherlands, England, Iceland, Japan, Malaysia, Mexico, Portugal, Singapore, India and Mozambique.

In carrying out the statistical analysis, the following independent variables were considered: sex, age, geographical location and scientific field. The dependent variables were related to the only item of the inquiry, as this had the unique goal of identifying which ideals scientists have and also to understand the possible role those ideals play. The answers were grouped according to three major categories: 1) collective ideals; 2) individual ideals; 3) without ideals. The first and the second categories are divided into three sub-categories: i) ideals related to science as a way of knowledge; ii) ideals related to science as a job; iii) ideals related to teaching. It is worth highlighting that the goal of this categorization is to identify contemporary ideals within contemporary science. The obtained data was analysed using SPSS.

Concerning the respondents, the following profile was obtained: the average age of all respondents was 41.7 years; the youngest individual was 24 years old and the oldest 70. The shunting line standard is approximately 8.4 and the variance is almost 70.

To the question: "What is your greatest ideal as a scientist/researcher?", comparing the groups of natural sciences and social sciences and humanities, it can be observed that the biggest disparities between them are found in what concerns the collective and individual ideals that are related to knowledge. The most immediate difference is that within the group of natural sciences there is a lesser notion of common good, which may be explained by the fact that current knowledge is too fragmented, just as described by authors like Edgar Morin, Peter Wagner, among others. On the other hand, the greater prevalence of individual ideals within the natural sciences, as well as with the ideals related to the job itself. It was also found that, in the case of the natural sciences, 31.6% of the respondents stated that they had no ideals. At first glance these results seem surprising as the majority of scientists did not have any ideals. Nevertheless, these findings ask for a more in depth analysis and that is what was done using a particular case: the scientific community of Catalonia. The aim was to analyse how the personal networks of researchers shape their ideals and if these play any role in the knowledge creation of academics.

2.5 Social network analysis

Social network analysis is a suitable method to test the impact of social and structural factors over processes, thus, it is a proper methodology to elucidate the relational context in which knowledge is created, by making visible the formal structures of groups and their meaning. Network research has shown that the interconnectedness of relationships between actors can influence behaviours through providing opportunities and inserting constraints [Brass et al., 2004].

In the particular case of studying knowledge creation, by focusing the relations and interactions among individuals, SNA is particularly appropriate for examining teams and collaboration networks (which fall between the micro-level of individual's context and the meso level of organizational features). SNA is also very useful to draw out precise linkages between agents engaged in collaborative activities as it reveals the structure of informal relationships amongst members and how such relations enhance or hinder certain types of activities that are crucial for the flow of resources or information or assets. In the particular case of this project, SNA offers an additional advantage of getting empirical data to fill in the schemes of the funds of knowledge; as individuals don't think in terms of network structures, when we show them the structure of the patterns of their relationships that allows a deeper inquiry into the meaning and content of those relationships.

In fact, one of the major advantages of SNA over other methodologies, including qualitative, is that it allows to systematically analyse the patterns of interactions among people in various types of social contexts and to quantitatively compare the structural features of such informal social networks

The analysis is based on the personal networks of a sample of 32 academic staff members of 4 institutions of the HE system in Catalonia, Spain: the Autonomous University of Barcelona (UAB, n = 21), the University of Barcelona (UB, n=2), the National Scientific Council of Research in Catalonia (CSIC, n = 5), and the Catalan Institute for Research and Advanced Studies (ICREA; n = 4). These networks were retrieved between May and August of 2013. The analysis of these networks aimed at exploring the properties of the networks that relate to the creation of knowledge in higher education.

For the selection of cases, we intended to maximize variation in terms of academic positions, duration of tenure, and academic discipline, although for various departments, multiple staff members were selected to better understand the importance of contextual effects. From the sample of 32 respondents, only 6 are female, due to a higher willingness of male researchers to participate in the survey. The departments were selected according to the level of knowledge production at the Autonomous University of Barcelona, i.e. considering indicators such as publications and research projects and the level of interdisciplinarity of the research fields and variety of academic positions.

Computer-assisted personal interviews were held using the software EgoNet (http://sourceforge.net/projects/egonet/). The first part of the interviews were structured, and consisted of the following modules. The first series of questions addressed the respondent. Study participants were asked about their motivations for research, main sources of information, working style, work environment, career, gender, and their use of social media networks for research purposes. Then, a set of 9 name generators was used to delineate the personal knowledge networks of the respondents. Specifically, respondents were asked who influenced them, with whom they collaborated, with whom they discussed research and career related issues, from whom they received social support, and who were their bosses, which doctoral students they currently supervised, with whom they had conflicts, and who they would go to for advice. For each name generator, respondents (egos) were allowed to nominate as many network members (alters) as they wished. Subsequently, questions about alters (name interpreters) and questions about the relationships among alters (name interconnectors) were posed, to measure network composition and structure.

Once the structured part of the interview was finished, a semi-structured part was held with two aims: first, to understand how respondents experience their knowledge creation processes, and second, to interpret the relational structures delineated quantitatively and their meaning for knowledge creation processes. The qualitative interviews focused on the respondent's perceptions on the networks of relations (attributes of the nodes), on the relational attributes (values, friendship, knowledge, information; active resistance) and on relational-based and network-based knowledge on the following topics: disciplinary differences; evolution of collaboration networks, academic knowledge and practical knowledge, mechanisms of exclusion and inclusion, criteria for research topics and curricula delineation.

The qualitative analysis of the personal knowledge networks of 32 academics of four HEIs in Catalonia, Spain [Ribeiro & Lubbers, forthcoming] showed a high correspondence between the personal knowledge networks and a permanent structural aspect of knowledge networks that is based on the role played by values on knowledge creation. It is to this personal networks that the respondents stick to no matter what the conjectural fluctuation (academic position shifts, criteria defined by journals, etc.) in their trajectory is. However, this is more due to the personal values attributed to specific ties than to dyadic similarity in other dimensions. Thus, scientific and personal values tend to be more appreciated as drivers for strongest ties than collaboration or expertise.

However, and not surprisingly, these values and ideals were more highlighted by older researchers than by younger researchers. Scientific generosity, openness, guidance, true friendship, companionship and purposeful research were the main values named by the informants. The emphasis on personal values and attitudes towards knowledge is the common denominator that explains the absent relation between similarity (in discipline and position) and individual knowledge creation and the positive relation between tie strength and the importance for knowledge creation. Furthermore, this emphasis on values often adopts a prescriptive stance on the research topics selected by the researchers and on their decisions regarding networking with other researchers. Those values are like an under the guise relational and academic code. This validation by means of personal values as a personal or rhetorical stunt often does not have any correspondence with the institutions or departments as these focuses primarily on competition, labour conditions and other quantifiable matters. These latter aspects also prove to be, on the other hand, fertile ground for this role played by personal values in the backbone of researcher's trajectories as they are what give stability to the researcher, even if he cannot research the topics he would like to.

An emphasis on values and ideals raises the question of whether researchers are more vulnerable to the impact of the weight of managerial prerogatives that dominate today's universities. Are researchers more subjected to value manipulation, since the emanations of the academic capitalism system and its implications on researchers are not attributable to one of the actors involved, but rather to a feedback loop that pushes forth cultural adaptation by permeation? To completely answer these questions is out of the scope of this paper, but it is important to notice that ideals and values become subjectified objects that afford layers of professional meaning and significance and undertake symbolic and actual confrontations that have never been intended or even conceived by their creators. The other side of this is raised by the new features of modern education (e.g.: the MOOCs and the increasing technologisation of teaching) that leads to the reciprocal objectification of subjects. As some literatures show (e.g.: Redman [2005]), teachers in shaping or resisting the interactive character of technologies concede something of their own sense of their autonomy [Fawns, 2008]. This 'something of their own' should be, I argue, ideals and values that overcome any type of objectification and survive change. Ideals are, thus, a way to rise above a conventional research perspective which is unable to make the transition from dualistic analytic categories of subject and object to the familiar complexity of a surrounding world of teacher education and of higher education. In addition, in any scientific endeavour or in any educational institution structural relationships must have a certain degree of duration and permanence. When these structural relations are embedded in useful and practical human and transdisciplinary ideals, they become the permanence in every quest for knowledge in face of any conjunctural fluctuation. Structural relations and scientific inquiries embedded in ideals are comparable to the circulatory system of an organism, in which liquids move faster or slower, but those moves don't imply a deep modification of the veins. That is the permanent structure that entails the circulation and, analogously, that structure serves as stable support to the continuous agitation and variability of scientific research and educational practices. This permanence dimension provided by ideals and values plays a crucial role in the characterization of the structural (as shown in the analysis of the knowledge networks mentioned in this work) and complements past and current models of educational models (see table 2.1).

Every time of crisis demands a double strategy to act reasonably. On one hand, one should accept the past lessons to make the most of the crisis and retrieve from it only what has to be kept. On the other hand, it urges the need and the emergence of a (re)newal to move civilization into the next step of evolution. Thus, I am not suggesting that we should reject all that has come with the conventional models based on the first level since its techniques and its partial truths are useful as long as they are applied into an adequate setting. At the same time, the double strategy that I propose with the predominance of ideals as drivers of education and scientific research requires the joint collaboration among ecologists, political scientists, sociologists, philosophers in order to produce solutions and ideas useful to the changes in all social levels with the proposed ideals as a starting point.

Level	Variables	Model
1. Techno economic	Micro-macro economy	Conventional for profit institutions
2. Socio-political	Quality of life, basic needs, groups, classes, institutions and power.	Broaden conventional; reformist, dependent model.
3.Ideals (axiological)	**Environmental limitations** **System of values and ideals**	**Ecodevelopment** **New development**

Table 2.1: Foundations of educational models

2.6 Conclusions

This paper has explored a way in which higher education can and should be engaged in an ideal grounded pedagogy and teaching. In order to do that, first, we explored which ideals researchers actually have and, second, how they shape researchers' knowledge networks.

I recognise that this ideal ground may seem to have a base of sand, but for the theoretical foundations and for the urgent and current need and potential they are the means to inform any quest that attempts to mix the contributions of sciences and humanities to improve higher education teaching and scientific research. To build a new society and to build ideals into reality demands a new language that cannot be detached from those same ideals. Everything, including technology can be used in a more humane way, in a more ideal way. Therefore, this is not an utopic proposal. It is utopic to think that the current models of development can continue to exist against all the limitations that we have (as the current crisis shows). To propose another basis as I do here – even with purposively less academic language – aims at suggesting different reflections and avoiding mental colonisations. However, for this to be possible science and humanities as science matter too.

References

Baert, P. [2012] "Positioning Theory and Intellectual Interventions", *Journal for the Theory of Social Behaviour* 42, 304-324.

Eyal, G., & Buchholz, L., [2010] "From the Sociology of Intellectuals to the Sociology of Interventions", *Annual Review of Sociology* 36, 117-137.

Edmunson, M. [2013] *Why teach? In defense of a real education.* (Bloomsbury, London).

Lamont, M. [2010] *How professors think.* (Harvard University Press, Cambridge, MA).

Pennac, D. [2008] *Mal de escuela* (Random House Mondadori, Barcelona).

Ribeiro, F. M. & Lubbers, M. [forthcoming] "Social networks and knowledge creation in higher education: the role of similarity and tie strength". *Procedia – Social and Behavioral Sciences.*

3

Science, Culture and 'Anti-science'

Jean-Patrick Connerade

Science today has sometimes a strange reputation of not really forming part of what ordinary European citizens call 'culture'. This peculiar view, while it is, of course, totally unjustified, nonetheless conveys a real difficulty. Scientists, over the past fifty years or so, have (it should be admitted) completely bungled their public image. As a consequence, they find themselves out on a limb, and there is now a serious need to reconstruct the relationship between science and other forms of high culture. This could be at the root of the whole 'anti-science' issue. Recently, it has been recognized that the problem exists. Indeed, CP Snow had brought up a related issue long ago, but researchers took little notice at the time, as he tended to blame artists and writers for their ignorance of science. Since then, the response has been at best clumsy, and demonstrates that, indeed, the relationship to other branches of culture has become somewhat distended. Basically, researchers have reacted to public hostility by arguing that science is relevant to the European citizen in terms of value added, where 'value' is the financial return on investment which should, in future, fund our expensive social welfare system in the 'knowledge society'. This is not really what the European citizen wants to hear, and he can probably guess it is a bit of a pipe dream. The proverbial *man in the Clapham omnibus* is fully aware of the difficulties of the modern world. He knows about competitivity and all questions surrounding the great economic shift towards the Pacific Rim. Economics, quite rightly, is not what colours his judgement about culture. What he really wonders is where modern science stands with respect to more fundamental values. Is it inside or outside our heritage and, if inside, how does it relate to our cultural identity? Until the question is answered clearly, the European citizen will continue to harbour doubts. This issue is illustrated by discussing the relationship between Science and Poetry, and how misunderstandings between them, born in the Romantic period, gave birth to the ecological movement. Harmony between the two, it is argued, will prove essential for the future of European science.

3.1 Introduction

Science today has sometimes a strange reputation of not belonging to what ordinary European citizens call 'culture'. This peculiar view is, of course, totally unjustified, but conveys a real difficulty. Scientists, over the past fifty years or so, have messed up their public image. As a result, they find themselves out on a limb, and there is now a serious need to reconstruct the relationship between science and other forms of creativity. This could well lie at the root of the whole 'anti-science' issue.

3.2 The 'Two Cultures' revisited

Fifty years ago, the novelist CP Snow brought up a similar question, but researchers took little notice at the time, as he mainly laid the blame on artists and writers for their ignorance of science [Snow 2012 (1959)]. Also, researchers were prosperous and successful, and saw their future as assured. Since then, the response to this issue has been at best clumsy from both sides. It demonstrates that, indeed, the relationship between science and other branches of culture is fraught.

To quote a recent statement:

"I would also like to convey that science is a fundamental part of our culture in Europe, it's exciting, it's full of opportunity and also fun...We can remind ourselves that Europe, through science, engineering, technology and philosophy, invented the modern world and it's also in our power to invent the future." said Prof Anne Glower [2014].

It isn't clear what she meant by "philosophy", but clearly, she needed it to justify the word "culture" without feeling guilty.

In reality, there is no need today for any misunderstandings. In fact, the digital age pervades the artistic culture of our times. Its methods are used everywhere, from music to the graphical arts, and even in literature. In fact, the internet and the personal computer have liberated the creative spirit of many modern artists. The only problem is recognition: these advances are not attributed to scientific research, their obvious origin. Just like glass or china, they have become props of daily life.

3.3 The 'economic' trap

Basically, researchers respond to public indifference or hostility by arguing that science is relevant to the European citizen in terms of 'value added', where 'value' is a financial return which should, in future, fund our expensive welfare system in the 'knowledge society'. If one follows this line, science must be practised because it is 'good for the economy' That may be true, but is far from enough. How can we be sure the future is better than the past?

Moreover, it is not what the European citizen wants to hear. He has become bored with economists bearing bad news, and he can probably guess that much of the talk about science working sudden economic miracles is a bit of a pipe dream. The proverbial *man in the Clapham omnibus* is aware of the complexity of the modern world. He hears every day about competitivity and the great economic shift to the Pacific rim. Economics, quite rightly, will not colour his judgements about culture. In fact, culture is based on values. Does science fit within this intellectual framework?

What he really wonders is where modern science stands with respect to more fundamental values. Is it inside or outside the mainstream cultural heritage and how does it relate to a European identity? Until the question is answered clearly, the European citizen continues to harbour doubts. He needs tangible examples, genuine role models, and wants to see how what he calls culture can possibly relate to what researchers call *science*.

3.4 The origin of 'Anti-science'

Against this background, the apparent vacuum between scientific and artistic cultures favours the growth of 'anti-science' movements. I suggest it is interesting to consider exactly where they come from, because their origin contains many lessons. They did not suddenly spring up in the USA, as some would believe. Interestingly, the scientists themselves have done much to help them grow. As I will stress, poets are closely associated with their birth.

The heyday of 'Anti-science' is not today, but in two moments of European cultural history: the Renaissance and Romantic periods. Starting with the European Romantics: we have the triangle Byron, Goethe and Mary Shelley, all inter-connected, through Faust, the Vampyre, the Sorcerer's Apprentice and Frankenstein. In these tales, they lay the basis of 'anti-science'. The poet Wordsworth further introduced a new theme: pollution through the burning of coal.

In the interest of brevity, I cannot tell the full tale of how these great writers came to their interwoven stories which all in some way involve man tampering with nature. What I want to emphasize, however, is that they were all in touch with each other and had a common source of inspiration, which led them to invent the modern gothic novel as a reaction against the industrial revolution.

3.5 The 'gothic' view of science

Science appears in this context as a form of modern devilry which releases destructive forces held in check by nature. It is actually a very fascinating side of Byron's romanticism, a reactionary view rooted in the rejection of infamous 'British' modernity. Byron's aristocratic tastes attracted him to Greco-Roman Antiquities and to the glories of Napoleon, but were hostile to trade and to the mercantile Society, ushered in by the Industrial Revolution born in the UK.

Two of his personal friends, Goethe and Mary Shelley, joined this Romantic conspiracy against science, and participated in propagating the 'anti-science' myths. Both were far from ignorant about science, but saw the literary opportunity. The poet Percy Bysshe Shelley also played a role. However, he was fascinated by science (especially chemistry) and indeed was held responsible for explosions and bad smells in the middle of the night at Oxford University before he was expelled.

Strictly speaking, the gothic novel is supposed to refer to some writings of Horace Walpole, who was the first to coin the phrase. In fact, however, the features we associate with it today have nothing much to do with stories penned by Walpole. They come rather from a notorious encounter, at the villa Diodati, near lake Geneva, between other famous poets and writers. Before we come to the plot, let me tell you something about the characters involved, set the scene and give you something of the flavour of the play.

At the villa Diodati were involved two front line Romantic poets, namely George Gordon, Lord Byron, and Percy Bysshe Shelley. Although they came subsequently to be considered as poets of equal stature, there was no comparison between their reputations at the time. Byron was writing a poem, the Childe Harold, which sold nearly two hundred and fifty thousand copies, a record at the time, and perhaps even to this day as far as poetry is concerned. He was at the height of his celebrity, and even once remarked about the success of his poetry: "I awoke and found myself famous". He had inherited a peerage, was rich and had been involved in a number of scandals, which seem to have driven him to flee from Britain. Naturally enough, with in addition a striking physique, he was a notable lady-killer. Despite the difference of reputation between them, it is remarkable that Byron immediately recognized and accepted Shelley as his equal in poetry. I humbly submit that great poets are often like that: they support each other, and we should learn from such a fine example.

Shelley, on the other hand, was little known at that time except perhaps for a variety of misdemeanours which had also led him to flee abroad. Most oddly, he had been expelled from Oxford University (rather hard to achieve), officially because he had written a tract in favour of atheism, but I suspect also because he was fascinated by chemistry to the extent of conducting experiments in his own rooms, causing sudden explosions and terrible smells in the middle of the night, which the other students complained about. He escaped to London, where he lived for a time, married his first wife, fathered children and wrote some of the earliest treatises on socialist philosophy (including one against wedlock). So far, still a bit ordinary, you might say, but, just as the treatises were being published, he ran off, not just with one, but actually with both the daughters of William Godwin, his own publisher. These daughters are rather important, so a word about them.

The first in order of importance is Mary Godwin, who married Shelley after the death of his first wife. The second (actually only her half-sister) was Mary Jane Clairmont, who may well have had her own affair with Shelley, but then became one of Byron's many mistresses and the mother of his daughter Allegra. Mary Shelley is of course very famous today as the author of Frankenstein, or the Modern Prometheus, one of the most important anti-science books in Western literature, but in truth, the plot of Frankenstein was conceived between her, Byron and Shelley, as she herself made clear, as a result of their crucial encounter at the villa Diodati.

The meeting included another interesting character, John Polidori, whose father had been personal secretary to the great Italian writer Alfieri. Byron was familiar with Alfieri's work and, although he originally hired his son as a private doctor, soon placed Polidori in a similar position to his father, as kind of personal secretary. Polidori was paid a handsome sum by the London publisher John Murray for an account of Byron's activities abroad, which understandably irritated Byron when he discovered it.

To the cast of characters, I would add one 'absent friend', namely Johann Wolfgang von Goethe, whose influence is essential. Byron did not know Goethe personally, but corresponded with him through friends as he knew no German, and we know that he admired his work through translations. Byron had travelled all over Italy with one of Goethe's books as his guide. Conversely, Goethe greatly admired Byron and was instrumental in promoting Byron's Don Juan in Germany, in particular at the Leipzig book fair, where it won the prize as the best book of the year. I will explain later why Goethe is so important to the story.

Now, consider the background. Great Britain, at the time, had seen the Industrial Revolution and was hard at work burning coal and becoming rich through highly polluting industries. Another famous poet, William Wordsworth, whom Byron much disliked, moved away from the grime of Manchester to the beauties of the Lake District. The characters involved in the Diodati episode knew all about the horrors of early manufacturing industry, but also knew that Great Britain was rich at the time precisely because of its industrial advantage. So it is quite natural that a love-hate relationship with science appeared amongst British Romantic poets a long time before other Europeans became aware of high profits and industrial wastelands. Consequently, the anti-science strand of Romanticism (we could call it gothicity) remained a peculiarly British phenomenon, and found no echo for a long time elsewhere.

Now, to the plot. Bear in mind that Byron could not remain for a moment near a woman of any beauty without attempting to seduce her. His strategy with Mary Godwin was to try and separate her intellectually from Shelley (an amateur chemist, remember, and an admirer of science) by insisting on the diabolical side of innovation. Mary's father, Godwin, was a friend of Erasmus Darwin, the grandfather of the great Darwin, and she knew quite a lot about biology, and especially about observations and experiments on the origins of life. In fact, she knew a lot more than either Byron or Shelley about the matter. Although she resisted Byron physically, it is clear she was swept off her feet by the towering intellect of the great poet. In her Preface to the 1831 edition of Frankenstein, she says "… Many and long were the conversations between Lord Byron and Shelley, to which I was a devout but silent listener. During one of these, various philosophical doctrines were discussed, and among others the nature and principle of life and whether there was any probability of its ever being discussed and communicated. They talked of the experiments of Dr. Darwin…" A commentator observes that, in all the novels Mary Shelley wrote after Frankenstein, there is always a male character resembling Byron. So, in this sense, there is little doubt his strategy worked.

Shelley was much more naïve about women and much more honest. He was so taken with storytelling, with the fantastic and the imaginary and the fully poetic side of things that he hardly noticed what Byron was really up to. For him, literature was the important outcome. When Mary recounted (not on the same day, as it happens, but somewhat later) her Frankenstein, he was insistent it was the best story of all they had produced. He helped her write it down and, indeed, wrote the introduction to the first edition himself, as Mary subsequently explained. He was oblivious of the fact that, in some sense, the story had been directed against him as a lover of science. Shelley was everything except small-minded.

At the villa Diodati, Byron, who had lived in South-Eastern Europe with an Albanian mistress, where he had picked up such tales, told probably the first vampire story of Western literature. He even wrote it down, but, as all his advances were resisted by Mary Shelley, lost interest, did not finish the manuscript and locked it away in a drawer.

There, it was discovered by Polidori (the villain of the piece) who completed Byron's unfinished manuscript and sent it to London for publication under his own name. There followed a huge quarrel, during which Byron, who was a very good shot, produced his duelling pistols and forced Polidori under threat of death to write a letter admitting that he had stolen and defaced his master's work.

Beyond the lovely tale, notice that the two main strands of the Gothic Novel, namely the vampire stories with their satanic undertones and the 'mad scientist' myth both appeared together, like the two sinister faces of a single coin. I think it is no exaggeration to say that any gothic literature down to this day always appeals to these two fundamental story lines, mixing them sometimes in greater or lesser degree.

Now, I turn to the influence of Goethe. Byron was of course aware of the Sorcerer's Apprentice (in German: Der Zauberlehrling a poem by Goethe, written in 1797) and also of the first version of Goethe's Faustus, published in 1808, one of the important works of the Sturm and Drang movement – a precursor to Romanticism in Germany. What is less known (but which Byron was fully aware of, being widely read) is that the myth of Faustus was actually the subject of a very remarkable play by the great Elizabethan poet Christopher Marlowe, a contemporary and friend of Shakespeare.

Although the legend of Faustus is possibly of Germanic origin, a certain Georgius Faustus Helmstetensis having been expelled from Ingolstadt for practising Astrology and Chiromancy, most of the story, including the relations of the pseudo scientist Doctor Faustus with the devil, is really due to the fertile imagination of Marlowe, who was himself a notable rogue and was accused of atheism and devil-worship in his own time. A troupe known as the Admiral's Men performed Doctor Faustus twenty-five times between 1594 and 1597 in London. It had a powerful effect on the first audiences. It is recounted by an early critic who wrote against the play and wanted it banned in 1632 that actual devils once appeared on the stage during a performance of Faustus, 'to the great amazement of both actors and spectators'. Some spectators were allegedly driven mad, 'distracted with that fearful sight'. In his Brief Lives, the chronicler John Aubrey recounts that the lead actor of the Admiral's Men repented and devoted his last years to charitable activities, such as founding Dulwich College, so impressed was he by the appearance of real devils on the stage…

If we try to understand where these inventions originate, we are inevitably led back to the clash between the Humanist philosophers and the theologians of the Middle Ages on the question: can knowledge can be acquired directly by observation of the world, or is it entirely contained within the Scriptures? This debate has been revived in recent times by the Creationists opposed to Darwin's theory, so we see an amazing continuity of debate in Western civilisation. If we refer to the Dominican Fra Girolamo Savonarola's great Sermon against the poets (de rationae poeticae artis published in Pescia in 1482, and again in Venice in 1511), we are led back yet further. Initially, Savonarola was attracted to Humanism and to the ideas of Giovanni Pico della Mirandola, but then turned against them. From his sermon, it is clear he knows all about the subject, but will have nothing to do with the new ideas. What he stresses is that everything to do with knowledge is already in the Holy Texts, that the world was created by the Almighty for man and that tampering with it, claiming knowledge acquired by other channels, or writing unnecessary poetry are all harmful activities and should be banned. In fact, he takes us right back to Plato's Republic and to arguments used by the Greek philosopher to forbid the pursuit of poetry. This is how Savonarola, who acquired a great popular following in Florence and even impressed Michelangelo, was able to organise the destruction of works of art and the burning of books in the middle of the Italian Renaissance.

It is clear that the arguments used by Savonarola are very close to the theme of Doctor Faustus, and that this theme, if we remove from it some of the more dated theological vocabulary, comes very close to Frankenstein and indeed to some quite contemporary views. Consider the following slightly modernised version of Savonarola's argument: 'Nature has provided us with what we should consider to be an ideal planet. Indeed, it is so perfect that there is no need to change it. To the contrary, we should avoid any kind of change, keep it exactly as it was made for us, preserve all the species living on it, and hopefully repair the damage humans have already done, since we all are part of a harmoniously created system, whose equilibrium must not be disrupted'. Behind the apparent wisdom of these words, we find the Romantic philosophy contained in Frankenstein; in the Sorcerer's Apprentice and in the Tragical History of the Life and Death of Doctor Faustus.

Again, Savonarola was not at all ignorant about humanism. Indeed, he had initially been attracted to it. Similarly, Goethe was a very competent scientist in his own time. Mary Shelley was the daughter of the philosopher Godwin, a personal friend of Darwin's grandfather Erasmus (himself a scientist). She knew a lot about fundamental advances made in biology. Erasmus Darwin experimented on the connection between life and electricity. He was also intrigued by the origin of species.

Without in any way denying the importance of preserving our planet from the harmful excesses of human activity, it is important to understand where the ideas underpinning the ecological movement originate, how they have been used in literature and what they actually imply. The role of poets in shaping them and giving them new life has been essential. What is also important to appreciate is that, even if the timescales can seem very long to us, we live in an ever-changing world: nature as we know it is evolving. Left to themselves, species come and go, planets change, stars age and worlds disappear. The nature we see about us is really due to millennia of agriculture, while original forests, full of tigers and other wild beasts were hardly nice places to live in. To humans, the realisation that the world is moving on is sometimes distressing, and to poets especially so, because eternal stability has something to do with glory and becoming famous for ever, even if nobody might read our work during our own lifetimes. We poets, ever since Horace, have always preferred to consider that Time would in the end be on our side, and science brings no comfort in that respect. To quote another poet who was very conscious of the Industrial Revolution and its dangers, Edgar Allan Poe:

Science! true daughter of Old Time thou art!
Who alterest all things with thy peering eyes.
Why preyest thou thus upon the poet's heart,
Vulture, whose wings are dull realities?
How should he love thee? or how deem thee wise,

Who wouldst not leave him in his wandering
To seek for treasure in the jewelled skies,
Albeit he soared with an undaunted wing?
Hast thou not dragged Diana from her car?
And driven the Hamadryad from the wood
To seek a shelter in some happier star?
Hast thou not torn the Naiad from her flood,
The Elfin from the green grass, and from me
The summer dream beneath the tamarind tree?

This question of the finiteness of time (or the end of Time) brings me perhaps to the last issue in which theology does shape the public perception of science, namely the Biblical Armageddon which precedes the Day of Judgement in the Apocalypse. Although there is much talk of eternity in religious texts, there is also an End of the World, which has not been a favourite theme of poets, I think, until the Cold War introduced and generalised the notion of impending nuclear doom. Again, the poets rose to the issue and attempted to picture the scene. Since I have not had time to quote much modern poetry, let me depict Armageddon through the words of Czeslaw Milosz [1980] in his *Child of Europe*

As we fled from the city in flames,
Along the first country path, looking back behind us
I said: Let grass one day cover the marks we made.
Let the wild prophets be silent in the fire
And let the dead recount to the dead what came to pass…

It is clear that poets can easily pick up all the bad news, but I would like to end on a much less sombre note, to show you that poets are also perfectly capable of restoring our confidence in progress. And since we have spent a lot of time with the Romantics, let me give you a French Romantic poet for the final note. Alfred de Vigny, in his famous poem *La bouteille à la mer*, gives us a metaphor well worth pondering. He describes a ship at sea caught up in a storm with little or no hope of survival, carrying something more precious than any cargo: scientific measurements, maps and information about new worlds which have been discovered on the way. The ship may be lost, but the precious intellectual cargo must be saved:

> Son sacrifice est fait; mais il faut que la terre
> Recueille du travail le pieux monument.
> C'est le journal savant, le calcul solitaire,
> Plus rare que la perle et que le diamant;
> C'est la carte des flots faite dans la tempête,
> La carte de l'écueil qui va briser sa tête:
> Aux voyageurs futurs sublime testament.

The captain puts the information in a sturdy bottle (appropriately enough: a Champagne bottle) corks it and throws it out to sea before the ship sinks. The wind and the currents bring the bottle back to France, where a fisherman finds it and wonders what the content can be. He takes it to an expert, who explains it thus:

> Quel est cet élixir? Pêcheur, c'est la science,
> C'est l'élixir divin que boivent les esprits,
> Trésor de la pensée et de l'expérience;
> Et si tes lourds filets, ô pêcheur, avaient pris
> L'or qui toujours serpente aux veines du Mexique,
> Les diamants de l'Inde et les perles d'Afrique,
> Ton labeur de ce jour aurait eu moins de prix.

The role of Poets in shaping opinion about science is key. In fact, scientists neglect poets at their peril, as this example shows. Consider the role of Faust. Actually, Goethe's Play was inspired by the Tragical Historie of Doctor Faustus by Christopher Marlowe, a friend of Shakespeare. Faustus himself was the incarnation of arguments advanced by the Renaissance theologians hostile to humanism.

The fact is simply that such literary texts play on two fundamental issues: the first is a semi-religious concern about not tampering with Creation, and the second is fear of the unknown. These great poets and writers have stated their case with considerable force, but where are the refutations? Scientists have not done their job: the public awaits eloquent and simple responses. It is not enough to consider such important texts as merely ignorant.

3.6 Why the Poets?

It is absolutely essential for the future of scientific research in Europe to involve, not only philosophers (themselves half-scientists), but also poets in rebuilding the image of the researcher as an honest character, who is not working against the future of mankind. This simple message cannot merely be carried by scientists, because they are the interested party. But it can be carried by poets, provided they become convinced it is true.

For this reason, the European Academy of Science, Arts and Literature (EASAL) supports, within the ESOF Forum of Euroscience, a special event under the title 'Science meets Poetry'. It brings together some thirty to forty poets, in close harmony with a major scientific gathering. We believe it has inspired poets about science, and also helps scientists to understand and respect poets as equals. We hope this will be a first step towards the reappraisal of science by a broader public.

Let me introduce this project. Many of you know ESOF, Europe's version of meetings such as AAAS in the USA. The Euroscience Open Forum was created in Stockholm in 2004 (while I was President of ES but the idea came from Professor Carl Sundberg of the Karolinska Institute in Sweden). It has been held every two years since (Munich 2006, Barcelona 2008, Turin 2010, Dublin 2012, Copenhagen 2014, Manchester 2016...) and attracts some 5000 participants (about 500 journalists). At the ESOF2006 in Munich, we first thought of inviting poets to celebrate the life and times of king Ludwig of Bavaria, who founded the University of Munich and who was something of an engineer, as well as being a poet. He was a pupil of the German chemist Justus Liebig, who thought very highly of him. Our experiment was very successful: it proved that poets and scientists mix very well. The book we published on the subject was also very popular. We decided to continue the experiment at ESOF2008 in Barcelona by introducing a one-day session entitled "Science meets Poetry" which was attended by the Nobel Laureate in chemistry Roald Hoffmann, himself a poet and today Professor of Humane Literature in New York. At ESOF2010 in Turin, we held open air sessions every evening in front of the Palazzo Carignano, to make 'S & P' accessible to all. At ESOF2012 in Dublin, the event was hosted in the Schrödinger Lecture theatre of Trinity College. It was attended by many local poets, including the late Seamus Heaney, Nobel laureate, reputedly the most widely read poet in the English language today. He was accompanied by Michael D. Higgins, a distinguished poet in his own right, President of the Republic of Ireland. ESOF2014 is scheduled from the 21st to the 26th of June in Copenhagen. The programme of the Science meets Poetry 4 event is already in preparation. Obviously, we will not miss the chance of being near Hamlet's palace! There are fascinating Danish connections involving the astronomer-poet Tycho Brahe, the science-inspired poet Edgar Allan Poe, the role of Hamlet and Ophelia in the poetry of Arthur Rimbaud, of Jules Laforgue and in the writings of Søren Kierkegaard, etc…

Such sessions are a unique feature of ESOF, and an important part of the European cultural landscape. They are open to all. Our aim is to change the view some have of science, and to change the public perception of poetry as well. We say to people of our times: come to Science meets Poetry 4 even if you are not yourself a scientist or a poet! The Proceedings are always published in book form after the event (those from Dublin are already available on the amazon website).[2]

3.7 Conclusion

More broadly, the conclusion has to be that a new humanism will be required in the XXI century to avoid regressive forces emerging under various disguises, ranging from economic and political to religious pressures. The Open Society is our inheritance, but we must fight to preserve it, or its advantages could all too easily disappear.

[2] See Connerade and McGovern [2013] and Connerade (Chaunes) [2014].

References

Connerade, J. P. & McGovern, I. (eds.) [2013] *Science meets Poetry 3: Proceedings from ESOF2012 in Dublin.* (Euroscience, Strasbourg).

Connerade, J. P. (Chaunes) (ed.) [2014] *Science meets Poetry 4: ESOF2014.* (Euroscience, Strasbourg)

Glower, A. [2014] *Address by Professor Anne Glower CBE, Chief Scientific Adviser to the European Commission,* at: https://cyber-zone.appspot.com/ec.europa.eu/archives/commission_2010-2014/president/chief-scientific-adviser/index_en.htm (accessed 7.09.2015).

Milosz, C. [1980] *Enfant d'Europe.* Translated into French from Hungarian by Monique Tschui and Jil Silberstein (with revision by the author) (Lausanne, Editions l'Age d'Homme). Extract retranslated into English by J.-P. Connerade for the present article

Snow, C. P. [2012] (1959) *The two cultures.* (Cambridge University Press, Cambridge).

4

Generation of 1911: A Case Study in Portugal

Maria Burguete

The generation of 1911, united by a common ideal of university and scientific research, formed a close-knit group with its own identity, in the contextualization of experimental medicine in Portugal, in the trail of a greater "scientificity" of this area of human knowledge. In the mid-19th century, clinical medicine in Lisbon confined to São José Hospital and Surgical School was performed in deplorable conditions. However, with the appearance of May Figueira (1829-1913) who studied Medicine in Paris and later did his PhD in Brussels, this situation changed completely, once his view upon clinical medicine was supported by the introduction of new apparatus, such as the microscope (an instrument magnifying objects by means of lenses so as to reveal details to naked eye), ophthalmoscope (an instrument for inspecting retina) and laryngoscope (a mirror apparatus for examining larynx). This new approach to clinical medicine based on experimental methodology leading to a more "scientific" approach of medical studies, was not well received by the great majority of their colleagues. This situation was natural considering that only 20% of the Portuguese population was literate. Nevertheless, in a great effort to achieve modernization in order to reach a European model of medicine from Germany, France and United Kingdom, a group of medical scientists, with no political affiliations characterized by their idealism on cultural and scientific accuracy, was capable of implementing this scientific methodology, launching the medical studies through an European dimension comparable to what was seen in Germany, France and UK in the same period. In this national context, the year of 1911 became a symbolic date for medicine in Portugal.

4.1 Introduction

The beginning of the XX century was launched into a new challenge by a few physicians that changed the face of medicine in Portugal: the birth of experimental medicine. In this paper we will discuss the role of the generation of 1911 in introducing experimental techniques of clinical diagnosis to the study of medicine. How this reform took place, and under what conditions this new era in medical studies - the experimental medicine – emerged, is the aim of this article.

The end of the XIX century in Lisbon was characterised by a school of medicine essentially clinical and performed at the old S. José hospital (Fig. 4.1) and Medical Surgery School in deplorable conditions, besides that hospital building.

Fig 4.1 - S. José Hospital Lisbon XIX century

In the transition to the XX century, medicine had experienced a deep revolution: the anatomic-clinical method from Germany was giving way to the new experimental medicine based on scientific methodology demanding creative activity and original research. Let us take a look to some personalities that were the "pierre de touche" or pillars of this new experimental era of medicine.

4.2 Development

4.2.1. Bernard and Ludwig: A New Version for Physiology

Claude Bernard (1813-1878), Fig. 4.2 and Carl Ludwig (1816—1895) Fig. 4.3, had shown a new version for physiology, a science of functions and phenomena of living organisms.

Fig. 4.2- Claude Bernard

This French physiologist taught at several major French institutions and was named a senator in 1869. He discovered the role of the pancreas in digestion, the glycogenic function of the liver in carbohydrate metabolism, and blood-supply regulation by the vasomotor nerves. His major contribution was to establish the principles of experimentation in the life sciences, including the need for a hypothesis. His concept of the internal environment of the organism led to the present understanding of homeostasis. Bernard also studied the effects of such poisons as carbon monoxide and curare. He was awarded the grand prize in physiology three times by the Academie des Sciences. Later on he turned to the philosophy of science and in 1865 published the famous "Introduction à la médecine expérimentale" (An Introduction to the Study of Experimental Medicine, 1927): this book discusses the importance of the constancy of the internal environment, refutes the notion of the 'vital force' to explain life, and emphasizes the need in planning experiments for a clear hypothesis to be stated, which may then be either proved or disproved. On the strength of this work he was elected to the French Academy in 1869. Claude Bernard's aim was to establish the use of the scientific method in medicine. He dismissed many previous misconceptions, took nothing for granted, and relied on experimentation. Unlike most of his contemporaries, he insisted that all living creatures were bound by the same laws as inanimate matter. In his own words:

> "When we meet a fact which contradicts a prevailing theory, we must accept the fact and abandon the theory, even when the theory is supported by great names and generally accepted".

Fig. 4.3 - Carl Ludwig

Ludwig's name is prominent in the history of physiology, and he had a large share in bringing about the change in the method of that science that took place in the middle of the 19th century with his friends Hermann Ludwig Ferdinand von Helmholtz, Ernst Wilhelm von Brücke, and Emil du Bois-Reymond, whom he met for the first time in Berlin in 1847. They all rejected the assumption that the phenomena of living animals depend on special biological laws and vital forces different from those that operate in the domain of inorganic nature, just like Claude Bernard (for further details please see Burguete [2014]).

4.2.2. German and French Influence: A New Vision

Therefore, in this new vision where vital forces were excluded, two assumptions can be made:

- Human organism is a chemical and metabolic world made up of complex mechanisms. This was the prelude of molecular biology and the conception of the human being as a biological organism.
- Microscopy made possible the analysis of tissues and cells (histology and citology) and the ethio-pathogenic concept was replaced by the discovery of many pathogenic microorganisms (Fig. 4.4 - Louis Pasteur and Fig. 4.5 - Robert Koch).

Fig. 4.4 - Louis Pasteur

Fig. 4.5 - Robert Koch

Louis Pasteur (1822-1895) was a French chemist and biologist who proved the germ theory of disease and invented the process of pasteurisation. The German doctor Robert Koch (1843-1910) is considered the founder of modern bacteriology. His discoveries made a significant contribution to the development of the first 'magic bullets' - chemicals developed to attack specific bacteria. For that achievement Koch was awarded a Nobel Prize in 1905.
The revolution of the human physical body was developed in parallel with the approach of psychological processes (the study of psyche or psychoanalysis) developed by Sigmund Freud (1856-1939), the founder of the International Association of Psychoanalysis in 1910.

4.2.3. The Dawn of Scientific Medicine in Portugal

The traditional medicine mostly made up of "clinical art" give way to the scientific medicine based on the basic experimental sciences, namely bacteriology and histology.

Now we are in position to look at the Portuguese case study and compare some situations, such as the case of the two physicians, Câmara Pestana

(1863-1899), Fig. 4.6 and Marck Athias (1875-1946), Fig. 4.7 considered the founding-fathers of the new Portuguese medicine, because since the very beginning of their experimental works, they both understand the importance of the new disciplines born from microscopy: bacteriology and histology.

Fig. 4.6 - Luis Câmara Pestana

Fig. 4.7 - Marck Athias

Two personalities should be mentioned from this period: Carlos May Figueira (1829-1913), Fig. 4.8 and Miguel Bombarda (1815-1910), Fig. 4.9.

Fig. 4.8 - Carlos May Figueira and his XIX century microscope.

Carlos May Figueira (1829-1913), a distinguished physician, was educated at Coimbra, Brussels, and Paris and went on to teach at the Escola Medico-Cirúrgica de Lisboa.

He is credited with introducing microscopy and the ophthalmoscope to Portugal in 1862, and played an important role in dealing with the cholera epidemic of 1856 and the yellow fever epidemic of 1857.

4.2.4. The Importance of Microscopy in Experimental Medicine

Early attempts to record medical microscopic images by photographic processes were done as soon as the technique was invented. In 1840, Alfred Donné, free professor at the *Faculté de medicine* in Paris, took some photomicrographies adapting a daguerreotype camera to the microscope. With his assistant the physicist Léon Foucault, Donné published in 1845 a medical course "*Cours de Microscopie*" to be used by his students. An Atlas was also published, accompanying this publication, containing microdaguerreotypes of human fluids and tissues.

Despite this pioneering work it would be another ten years before photomicrography became implemented as a tool in medical research and teaching; questions were still raised on the use of microscopy for medical diagnosis. The invention of the gelatine-dry plates in 1871 by the physician Richard Maddox, himself a keen microscopist, and improvements on optical equipment and artificial lights, led to an adequate use of photomicrography in medicine and by the end of the 1870s bacteria was being photographed.

Fig. 4.9 - Miguel Bombarda

Miguel Bombarda, director of Rilhafoles Hospital in Lisbon had created in 1892 and1897, both the Bacteriological Institute and Histology Laboratory, being these institutions, the main pillars of the experimental studies in medicine. Under the influence of these men two important laws in 1911 arose:

- In 22 February medical studies were reformulated.
- In 24 March two new universities and faculties emerged: medical schools from Lisbon and Porto became faculties with universitary status.

Of these two physicians, only May Figueira was an experimentalist; the other was essentially medical clinical doctor with a vision of future which lead to the emergence of the pillars of experimental medicine studies in Portugal.

4.2.5. Public Health in Portugal

By the end of the 19th century, microscopy was acknowledged in the country as an important tool to study diseases as we can see by the works done by the Portuguese bacteriologists Ricardo Jorge (1858-1939), Fig. 4.10 and Anibal Bettencourt (1868-1930), Fig. 4.11 who made photomicrography a central tool for documenting their medical diagnosis.

Fig. 4.10 - Ricardo Jorge

Fig. 4.11 - Anibal Bettencourt

Ricardo Jorge was pioneer of Public Health in Portugal and in 1885 published the famous book "Social Hygienic Conditions at Portuguese Nation" creating a new era in Public Health in Portugal similarly to what happened more than 40 years ago in UK with Edwin Chadwick (1800-1890) an English physician "Report on Sanitary Conditions of Working People at UK".

Anibal Bettencourt, physician and bacteriologist, was appointed director of Bacteriologic Institute after Câmara Pestana's death in 1899. He created the scientific bases for tropical medicine with the study of "Sleeping Disease" in Angola (tripanossomíase). In this picture Fig. 4.12, an oil painting from Columbano Bordalo Pinheiro, we can see the five physicians from Escola Médico-Cirúrgica de Lisboa in 1907:

Fig. 4.12 "Lentes da Escola Médico-Cirúrgica de Lisboa" (from left to right: Eduardo Mota; May Figueira, Sabino Coelho, Ricardo Jorge and Oliveira Feijão)

Therefore, the heirs of these institutions, Luís Câmara Pestana and Marck Athias, were given the opportunity to fund a new experimental medicine school in Lisbon. Marck Athias was then appointed director of the Histology Laboratory by Miguel Bombarda.

4.2.6. Experimental Medicine Institutions in Portugal, Germany and the United Kingdom

A pioneer in bacteriology in Portugal, Câmara Pestana, under the supervision of Sousa Martins, worked in Bacteriological Institute (Fig. 4.13) founded in 1892, helping patients receiving treatment against "rage" in perfect sanitary conditions as it was previously done in Paris. Therefore, all microbiologic analysis was performed to control treatments, and those treatments were also used as teaching material to practical medicine studies. After his premature death in 1899, his work was developed by his successor, Anibal Bettencourt, a man of great vision that change this Institute into a great laboratory of experimental medicine. Câmara Pestana followed the French model of Institute Pasteur in Paris (1888), where he complemented his bacteriologic studies. The Robert Koch Institute in Berlin was founded in 1889 and the Lister Institute in London in 1889 also followed the French model.

Fig. 4.13 - Câmara Pestana Bacteriological Institute

A beautiful collection of scientific equipment as well as a great library turned this beautiful building from the beginning of the XX century, a masterpiece of experimental medicine studies in Portugal.

The Robert Koch Institute (RKI), Fig. 4.14, is Germany's national Public Health Institute. It is the central institution of the Federal Government in the field of surveillance, control and prevention of diseases and hence also of application-oriented biomedical research.

In 1891 the British Institute for Preventive Medicine (BIPM) was incorporated. In 1893 the BIPM amalgamated with the College of State Medicine. In 1898 Lord Lister became Chairman of the Governing Body of the BIPM. In 1899 the BIPM became the Jenner Institute of Preventive Medicine. In 1903 the Institute became the Lister Institute of Preventive Medicine (LIPM), Fig. 4.15.

Fig. 4.14 - Robert Koch Institute Berlin

Fig. 4.15 - Lister Institute London

4.2.7. The 1911 Generation in Portugal

4.2.7.1 Marck Athias (1875-1946) and his research school

The first psychiatry hospital in Portugal was built in 1853 – Rilhafoles Hospital - in Lisbon by the Portuguese Queen, Dona Maria. However, only in 1896 was the first course of psychiatry established in the same place by the physician Miguel Bombarda, who was appointed director of the laboratory of Histology of that hospital. One year later in 1897, that place was named Miguel Bombarda Hospital. That institution has worked for almost 200 years until in 2011 its activity has come to an end.

Fig. 4.16 - Marck Athias and Miguel Bombarda Hospital

Marck Athias reached Lisbon in 1897, after the conclusion of his studies at the Faculty of Medicine in Paris where he was very influenced by the ideas and works in nervous histophysiology of Santiago Ramón y Cajal. Therefore, Marck Athias (Fig. 4.16) was welcomed in 1897 by Miguel Bombarda, with whom he shared the defence of the theory of Cajal's neuron. Marck Athias replaced Miguel Bombarda as director of histology at Miguel Bombarda hospital. Conditions were created to the research programme developed by Marck Athias' school starting with nervous histophysiology in 1897, an area in which Athias made the first scientific contributions in 1895, and continued until 1915.

Both Marck Athias and Câmara Pestana realized the importance of these new disciplines born from microscopy: histology and bacteriology. Athias and their disciples, Augusto Celestino da Costa (1884-1956), Abel Salazar (1889-1946) and Ferreira de Mira (1875- 1953) among others developed their training in experimental methods in several scientific fields such as histology, histophysiology, histopathology, physiology and physiological chemistry (see ref. Burguete [2013]). The disciples of Athias, particularly Celestino da Costa and Ferreira de Mira, echoed through their doctrinal writings the debates on issues that took place in Europe, especially in France, between 1870 and World War I. They also defended scientific dissemination and propaganda in society in general, the promotion of a scientific culture among researchers as a form of counterbalancing an excessive specialization, the scientific interchange between people and institutions and, finally, the creation of scientific societies and specialized publications.

The research programme developed by Marck Athias' school started with nervous histophysiology in 1897, an area in which Athias made the first scientific contributions in 1895, and continued until 1915. In 1905 Marck Athias joined the Bacteriological Institute with an original experimental work entitled "Nervous Cell Anatomy". This area expanded to include general histophysiology from 1905 onwards, branching into physiology and physiological chemistry from 1911. Therefore, under the influence of the masters, Aníbal Bettencourt and Marck Athias, a new vision of medicine was achieved through the evolution of a scientific and experimental research activity. These activities were supported by well-equipped and organized laboratories, where those masters could lead a new group of young researchers that followed the new experimental medicine, the so called "Generation of 1911".

4.2.7.2 Medical Sciences Institutions in Europe and USA

This was a unique phenomenon: at the end of the XIX century, Portugal had conquered a pioneer position within the world medical science. Let us take a look at the world picture concerning the birth of medical sciences institutions (Fig. 4.17-Fig. 2.23):

- The Pasteur Institute of Paris was founded by Louis Pasteur in 1887 - This worldwide biomedical research organization based in Paris was the first to isolate HIV, the virus that causes AIDS, in 1983.

Fig. 4.17 Pasteur Institute of Paris

- Bacteriological Institute of **Lisbon** in 1892/Câmara Pestana

Fig. 4.18 - Bacteriological Institute of Lisbon

- Laboratory of Histology of **Lisbon** in 1897 / Miguel Bombarda.

Fig. 4.19 - Laboratory of Histology of Lisbon

- Cajal Biology Laboratory of **Madrid** in 1901/Ramon Cajal

Fig. 4.20 - Cajal Biology Laboratory of Madrid

In this laboratory Cajal developed his neuron theory. This is the most ancient neuroscience institution in Spain.

- The Rockefeller Institute in **USA** for Medical Research (1901) became later Rockefeller University and was the first medical research Institute in USA linked to a university chartered to develop a scientific understanding of "the nature and causes of disease and the methods of its treatment, and to make knowledge relating to these various subjects available for the protection of the health of the public and the improved treatment of disease and injury."

Fig. 4.21 - The Rockefeller Institute

Fig 4.22 - Rockefeller University

On October 26, 1910, the new Rockefeller Institute Hospital admitted its first research participant, opening up a new era of biomedical investigation in which physicians were given the resources and encouragement to engage in fundamental studies in the hospital laboratories on the disease problems they dealt with on the wards of the hospital.

Kaiser Wilhelm Gesellschaft of **Berlin** was born in 1911 and later on 1948 became property of Max-Planck Society.

a) b)

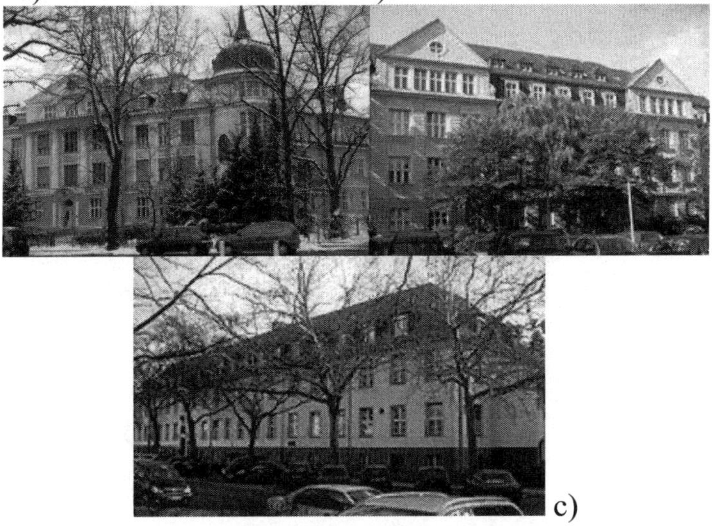

c)

a) Institute for Chemistry b) Institute for Biology c) Institute for Anthropology

Fig. 4.23 – Kaiser Wilhelm Gesellschaft of Berlin

(actually the Berlin Free University)

In the following slide we can have an overall picture of the whole situation concerning the first medical science institutions in Europe and USA:

Another important feature we must consider in the development of medical sciences in Portugal was the performance in 1906 of the XV International Congress of Medicine in Lisbon organized by Miguel Bombarda. This successful event mobilized more than 2000 participants from 27 countries all over the world, with 134 thematics of study and 500 communications or abstracts. More important even, was the publication in 1907 of all these works in more than 20 volumes of proceedings.

Finally, the challenge was settled and all Portuguese physicians from the generation of 1911 wanted to go to Europe (the world cultural centre in the XX century) in order to learn more about this new experimental medicine. This was the situation with the following personalities:

4.2.7.3 Small Biography of the Generation 1911 Group

Celestino da Costa (1884-1956): studied in Berlin with Hertwig, Krause and Retterer.

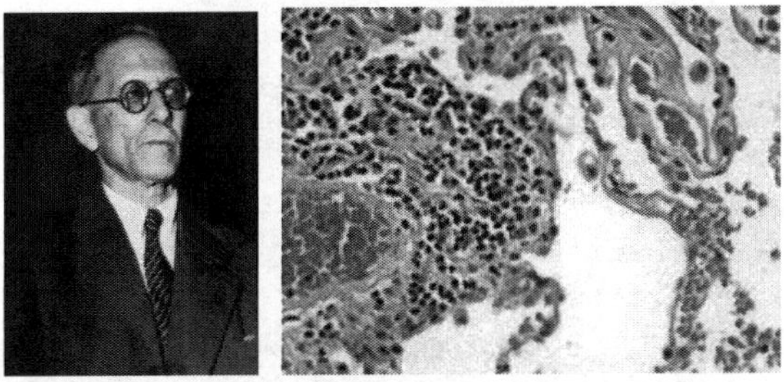

Fig. 4.24 – Celestino da Costa/lung histology slices.

He was the first embryologist in Portugal and Spain. The founder of **Institute of Histology and Embryology** of Faculty of Medicine from Lisbon University.

Sílvio Rebello (1879-1933): connected to Schmiedberg Laboratories in Strasbourg. Schmiedeberg, Johann Ernst Ostwald (1838-1921) took a year's leave in 1871 to work in Carl Ludwig's institute of physiology at Leipzig, where he sharpened his skills in medical research and made contacts with colleagues who would have important roles in future physiological investigations. Early in his tenure at Strasbourg, the authorities decided that Schmiedeberg would have an institute of pharmacology near the medical centre. The building, which was opened in 1887, was carefully planned by Schmiedeberg and provided laboratories for animal experimentation and chemical studies as well as offices, an auditorium, and a library. This institute attracted young scientists who became the foundation of the new profession. More than 150 pharmacologists were educated by Schmiedeberg, including more than thirty from countries outside Germany. His outstanding students included John J. Abel, Arthur R. Cushny, Hans H. Meyer, Torald H. Sollmann, and Sílvio Rebello all of whom had an important role in training the next generation of pharmacologists.

Fig. 4.25 - Sílvio Rebello

Founder of Pharmacology **Institute** introducing the physico-chemical basis of pharmacological sciences. This physician introduced the art of research particularly the importance of experimental design and execution of experiments.

Reynaldo dos Santos (1880-1970) studied in several European surgical centres from France, Germany, Italy and United Kingdom). Pioneer surgeon and original developer of diagnostic arterial visualization techniques and technologies. He worked with Werner Forssmann (Nobel prize in 1956) and in this book, "Peripheral Endovascular Interventions" we can see a picture from both scientists side by side and we can read the following on p.4 (see Fig. 2.26)

> … In their 1931 book they published 100 peripheral artheriograms and 300 lumbar aortograms encompassing pathological findings in various inflammatory conditions and tumors. This work paved the way for the future development of vascular visualization (see Fig. 2.26).

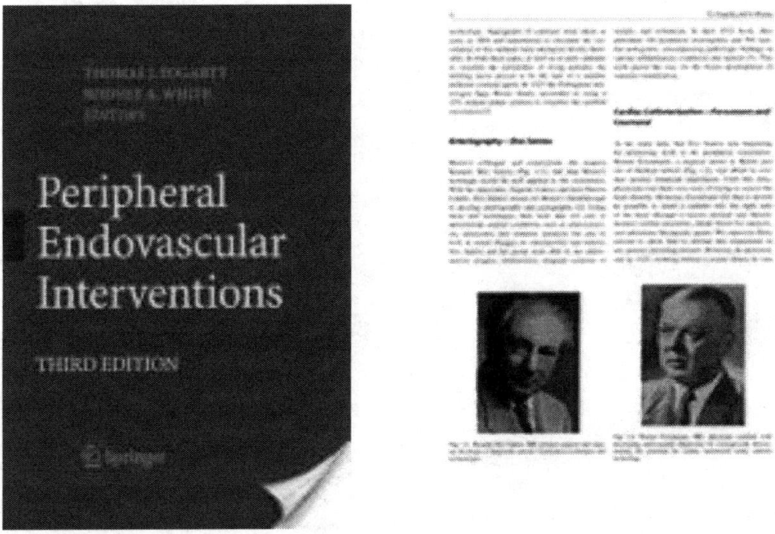

Fig. 4.26 – From left to right: Reynaldo dos Santos and Werner Forssmann

Reynaldo dos Santos, invented aortography (the first radiographic visualization of abdominal circulation in a living being), and arteriographia in 1929, culminating with the introduction in 1939 of the concept "Arterial conduction in semiology and therapeutics"- every drug/substance or instrument introduced into our organism is made up through arterial conduction.

Reynaldo dos Santos was a colleague of Egas Moniz (NP in 1949), who had been a disciple of Miguel Bombarda.

The history of cerebral angiography of Egas Moniz: (Fig. 2.27) had been recently exhibited in 2005 at Museu Nacional de Arte Antiga, in Lisbon (photo from Nina Szielasko).

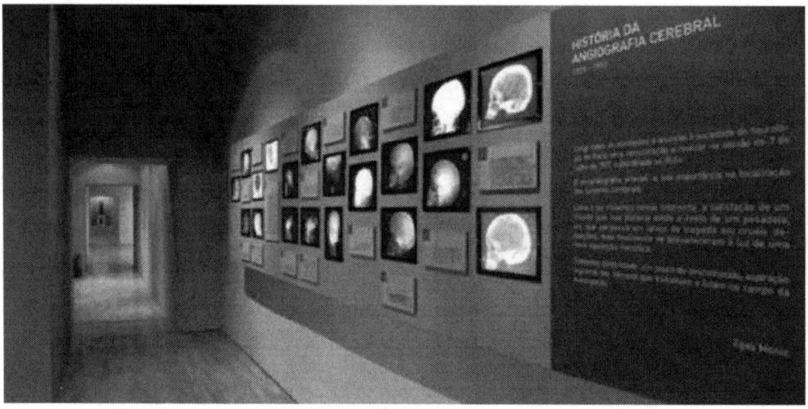

Fig. 4.27 - The history of cerebral angiography of Egas Moniz.
Exhibition at Museu Nacional de Arte Antiga, Portugal

João Alberto Pereira de Azevedo Neves (1877-1955) also studied anatomy and pathology in Germany (Hamburg, Berlin and Dresden)

Fig. 4.28 – João Alberto Azevedo Neves

In 1903 the Portuguese Government sent Professor Azevedo Neves to Copenhagen to study Finsen Method and Lupus Phototherapy and in the same year a book of 200 pages, "O método de Finsen para o Tratamento de Lupus Vulgar", was written as a detailed report about his scientific voyage to Copenhagen together with a small course of PhotoBiology, fully illustrated having Finsen Institut as the major actor, including a detailed description of laboratories and the working techniques. A similar laboratory was then implemented at S. José Hospital to treat patients with this disease. In 1907, Azevedo Neves together with Marck Athias were appointed to the Pasteur Institute of Lisbon where Finsen phototherapy was applied to the Lupus treatment. Therefore, the study and development of phototherapy methods have been implemented not only on private institutions but also at the service of Public Health.

Being also a prestigious pathologist he was appointed 1911 as Director of Pathologic Anatomy Institute (Instituto de Anatomia Patológica e Patologia Geral); later on he decided to move into the Legal Medicine Institute, an institution founded by himself also in 1911. In the figure below (p. 140) was taken from the book (vol. 3) *"Histórias da Luz e das Cores"* written in 2010 by Professor Dr. Luis Miguel Bernardo, director of Porto Science Museum of Porto University, where the work of Azevedo Neves is presented:

Em Portugal a fototerapia de Finsen não passou despercebida. Virgílio Machado, depois de ter visitado vários centros de saúde estrangeiros, fundou em 1903 o Instituto Médico Virgílio Machado, onde, para além de outras técnicas físicas de diagnóstico e terapia (raios X e electroterapia), era usada a fototerapia. O governo português enviou em missão especial a Copenhaga o professor João Alberto Pereira de Azevedo Neves (1877-1955) que, ao regressar a Portugal, montou no Hospital de S. José um serviço com aparelhagem de raios Finsen, tendo treinado a Dra. Sophia Quintino, que se especializou nesta área médica e publicou numerosos trabalhos. Azevedo Neves publicou após o seu regresso um folheto de 40 páginas intitulado *O Methodo de Finsen para o Tratamento do Lupus Vulgar* (1903) e, em 1904, um relatório da viagem de estudo com 158 páginas e o mesmo título. A primeira parte deste livro é um interessante compêndio de fotobiologia; na segunda parte, dedicada ao Instituto Finsen, o autor faz uma descrição pormenorizada das instalações, recursos e pessoal, bem como do laboratório e da clínica. Para terminar, Azevedo Neves dá informações sobre as capacidades técnicas fototerapêuticas de Hamburgo, Berlim e Dresde. Este livro mereceu o elogio do jornal científico francês *Le Radium*. Posteriormente, publicou ainda um relatório *Terapêutica do Lupus Vulgar* (1904), bem como um folheto resumindo uma conferência que fez sobre Niels R. Finsen (1905). Em 1907, Azevedo Neves ocupava juntamente com Mark Athias a direcção do Instituto Pasteur de Lisboa, onde existia uma secção de fotografia e se realizavam tratamentos pelo método de Finsen e por raios X. A prática e o estudo da fototerapia desenvolveram-se em Portugal tanto nas escolas médicas como nos hospitais e institutos médicos privados. Em 1912, o serviço de radiologia do Hospital Geral de Santo António, no Porto, estava equipado com um aparelho de arco voltaico Finsen-Reyn de 25 A e "carvões com alma de ferro, que teem a vantagem de augmentar o numero de raios azues e sobretudo violetas que são aproveitados no tratamento pelo methodo de Finsen." Nesta época, também na cidade do Porto, era criado pelo medico Jaime de Almeida um "Instituto Electrotherapeutico" para tratamento de doenças artríticas e nervosas

Fig. 4.29 – Description of Azevedo Neves work by the author Luis Bernardo

Francisco Gentil (1878-1964), Fig. 2.30 was pioneer at Oncology in Portugal, while organizing for the first time in 1911, a hospital for cancer patients. This task had been decided by the council of Lisbon Faculty of Medicine, founded in 1911.

Fig. 4.30 - Francisco Gentil and Oncology Hospital

Henrique de Vilhena (1879-1958), Fig 4.31 with was a professor of Anatomy at the School of Arts in Lisbon and together with Celestino da Costa he founded the Portuguese Society of Anatomy, where all anatomists from Lisbon, Porto and Coimbra were together.

Fig. 4.31 - Henrique de Vilhena

These young physicians had now new buildings to develop their experimental studies and proceed with their experimental medicine methodology already implemented by their followers and research group associated with them all over Europe. Fig. 4.32-a) and b) summarise all the medical institutions in Lisbon 1911 as well as the generation of 1911 involved in this adventure within the experimental medical sciences in Portugal:

a)

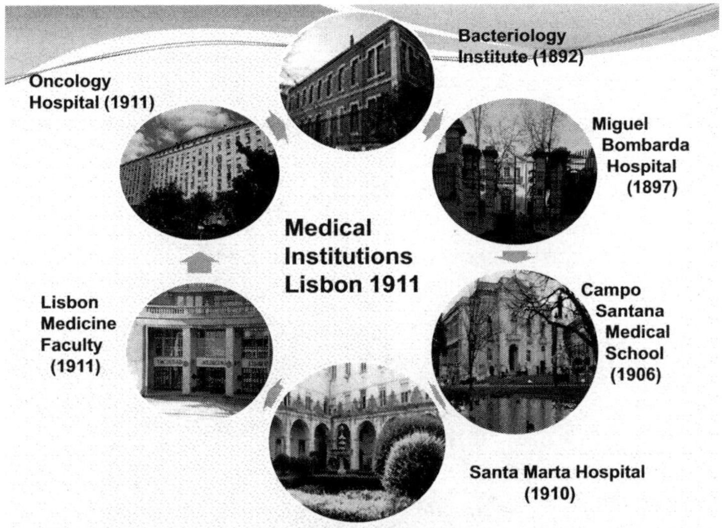

b)

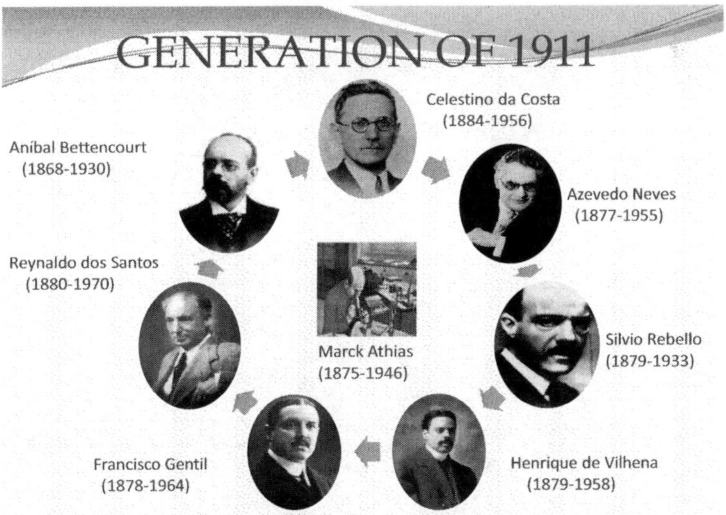

Fig. 4.32 – a) Medical Institutions and b) Generation of 1911
Campo Santana Medical School of Lisbon

In 1836 the School of Pharmacy was integrated into the Lisbon School of Surgery and only around 1910 was it incorporated into the faculty of Medicine of Lisbon University.

4.2.8. Medical Schools in Portugal and Experimental Medicine

Campo Santana Medical School was born in 1906 for the teaching of medical basic sciences while the clinic teaching was still performed at São José Hospital until 1910. Then, as the Santa Marta School Hospital it was given in 1910 to the Faculty of Medicine, this would be the new place where medical teaching will take place together with Campo Santana Medical School. According to the law of 22 February 1911, it was possible from then on to perform a deep revolution in all medical sciences methodology, introducing the experimental field of medicine, reforming all the old laboratories into new Institutes, where new experimental works were introduced.

For this purpose, Marck Athias was responsible for the selection of new apparatus and equipment to establish the Institute of Physiology in 1911 together with Ferreira de Mira and Joaquim Fontes. In this same year, the Institute of Pharmacology was established by Silvio Rebelo, who has brought to Portugal the knowledge how chemical substances interact with biological systems or Pharmacology. Celestino da Costa, also in 1911 established the Institute of Embryology and Histology with Pedro Chaves, Alfredo Ramalho and Luís Raposo. The Institute of Anatomy was funded by Henrique de Vilhena, Vitor Fontes and Barbosa Sueiro, in spite the maintenance of the old methodology. Finally, the Institute of Pathologic Anatomy and General Pathology also emerged in 1911 and Azevedo Neves used his influence to build an Institute of Legal Medicine next to the Faculty of Medicine.

Santa Marta Medical School Hospital

Santa Marta Medical School Hospital is located in the old monastery (Santa Marta Monastery) from the XVI century, converted into a Hospital for poor people in 1890 when a flu epidemic arose in Portugal. Later on, in 1903 it has become a property dependent on the Portuguese Government with the aim of treating venereal diseases. This hospital was similar to São José and both became the so called Lisbon Civil Hospitals. From 1910 on this place became the New Medical School of Surgery until 1953, when a new Universitary city emerged with all the faculties inserted in a whole city at Campo Grande. From 2013 on it will be transferred to the new "Hospital de Todos os Santos" at Parque da Bela Vista. In 1934 Santa Maria Hospital joined this group (Lisbon Civil Hospitals) through the hand of Professor Francisco Gentil.

In Santa Marta Medical School a new environment completely different from those of the Civil Hospitals of Lisbon emerged, once there was a straight connection with the Ministry of Education. Naturally, this connection gave this institution a considerable degree of independence and autonomy, therefore achieving the name of universitary clinics, each one possessing its own laboratory, archives and library for research purposes.

Faculty of Medicine University Lisbon

Fig. 4.33 – Faculty of Medicine University Lisbon

This new faculty founded in 1911 was now ready to perform its mission: the development of modern medicine with new physicians well prepared abroad and the old laboratories transformed into new institutes (see below), where real laboratories well equipped with new instrumentation could give raise to excellent work conditions. Once again, Marck Athias made his contribution by selecting the best equipment as well as the best scientific revues and books to every Institute so they could afford their own private and specialized library. This work was possible by travelling in 1909 to Paris, Gand, Brussels, Leipzig and Göttingen, in search for the best equipment and books. Just like Costa Simões did in 1866 while reforming the medicine course at Coimbra University [Burguete, 2010].

Physiology and Physiological Chemistry Institute
Founded by **Marck Athias**, Joaquim Fontes and Ferreira de Mira in 1910 this was an important research and learning centre. Later on, Athias also worked at Rocha Cabral Institute, Pasteur Institute of Lisbon and IPO (Portuguese Oncology Institute).

Institute of Pharmacology and General Therapeutics
Founded by **Sílvio Rebello** in 1910 together with Toscano Rico and Gomes da Costa, this place developed the study of Pharmacology where the biological action of substances started to be study in detail.

Institute of Histology and Embryology
Founded by **Celestino da Costa** in 1911 together with Pedro Roberto Chaves, Alfredo Magalhães Ramalho and Simões Raposo. Celestino da Costa, a university professor and a scientific researcher, he was the father of Embryology in Iberia having published all the didactic books needed to perform his experimental works, not only in Portuguese language but also in Spanish and French. His books were expanded to France, Belgium, Spain and South America.

Institute of Anatomy
Founded by **Henrique de Vilhena** (1879-1958), with Vitor Fontes (1893-1979) and Barbosa Soeiro (1894-1974).

Institute of Pathologic and General Anatomy
This Institute founded by António Pinto de Magalhães had a small period of life since its founder died soon after its opening. Only in 1936, when the great German pathologist Friedrich Wohlwill arrived Portugal as a German refugee, this institute developed a true method of clinic-anatomy together with **Pulido Valente** (1884-1963), Fig. 4.34:

Fig. 4.34 - Pulido Valente

4.3 Conclusion

In conclusion we can say that, an approach at the cell level and not only at the organ level, was achieved with physiological chemistry; a shift from the cellular to the molecular level was announced, within the scope of physiology (electric phenomena in nerves and muscles; muscle tension and muscle power; acoustic connections; optical questions and the visual ability of the eye), histology and bacteriology.

These fields of research led to the emergence of Public Health in Portugal with the possibility of the development of tropical medicine once Portugal was also Africa (Angola, Mozambique, South Africa, Guinea, Cabo Verde, Timor); Asia (Macau) and South America (Brazil).

The generation of 1911 developed a singular movement with the main scientific research ingredients:

Good working conditions including fine buildings and instruments, curiosity, hard work, network, knowledge, innovation and imagination! The right persons at the right places at the right moment!

The singular movement of generation 1911 was indeed a very special case study – an oasis within the national context – however, that sole movement alone, could not change the course of Portugal history.

There are also some other names connected to this movement of the "generation of 1911" such as: Egas Moniz (1874-1955); Gama Pinto (1853-1945); Eduardo Mota (1817-1912); Sabino Coelho (1853-1938), Oliveira Feijão (1850-1918), José António Serrano (1851-1904), Silva Amado (1846-1925), Curry Cabral (1844-1920), Custódio Cabeça (1866-1936), Alfredo da Costa (1859-1910), José Gentil (1870-1941), Ferraz de

Macedo (1838-1907), Bello Moraes (1868-1933), Moreira Júnior (1866-1952), Bettencourt Pita (1826-1907), Bettencourt Raposo (1853-1937), Sílvio Rachello (1879-1938), Salazar de Sousa (1871-1940), Carlos Tavares (1857-1913), Augusto Vasconcelos (1867-1951).

References

Alves, M. V. [2005] "O projecto Museu de Medicina da Faculdade de Medicina de Lisboa", *Revista da Faculdade de Medicina de Lisboa 3,* pp. 155-167.

da Costa, J. C. [1999] "A irreprimível vitalidade de Reynaldo dos Santos", in *1911-1999. O Ensino Médico em Lisboa no Início do Século. Sete Artistas Contemporâneos Evocam a Geração Médica de 1911* ed. Alves, M. V. (Fundação Calouste Gulbenkian, Lisboa) p. 139.

Património da Universidade de Lisboa [2011] Ciência e Arte. edições Tinta da China. Comemorações dos 100 anos da Universidade de Lisboa.

Serrano, J. A. (ed.) [1901] *Anuário da Escola Médico-Cirúrgica de Lisboa*, ano lectivo 1899/1900 (Imprensa Nacional, Lisboa).

Dias, J. P. S. [2011] "O Instituto Bacteriológico: espaço, instrumentos e memória da medicina laboratorial", in *Património da Universidade de Lisboa. Ciência e Arte,* eds. Lourenço, M. C. & Neto, M. J., (Universidade de Lisboa/Tinta da China, Lisboa) pp. 137-154.

Burguete, M. [2014] "Medical Studies in Coimbra 1911", in *All about Science: Philosophy, History, Sociology & Communication,* eds. Burguete, M. & Lam, L. (World Scientific, Singapore) pp. 193-208.

Burguete, M., Martins, D.R, & Fiolhais, C. [2013] "Evolução dos Estudos Médicos Em Coimbra no Século XIX – Contribuição das Ciências Físicas e Químicas", in *História da Ciência na Universidade de Coimbra 1772-1933* (Imprensa da Universidade de Coimbra, Coimbra) pp. 167-192.

Burguete, M. [2010] *Medical Studies at Coimbra in the XIX Century: faculty of Medicine Laboratories and Scientific Travelers*. (Lap Lambert Publications, Saarbrücken).

5

Remembering and Forgetting: A Visual Arts Research Practice in Scientific Laboratories

Maria Manuela Lopes

Artistic residencies in scientific laboratories are becoming common practices in the UK but are still fairly recent and innovative in Portugal. I take my experience in the artistic residencies in scientific domains and my arts-based PhD research in a neuroscience laboratory as a starting point to consider artistic encounters with distinct fields of biological and health research and practice. I explore my practice-based projects developed progressively through a body of interrelated installation artworks intertwined with the themes of memory. The personal practice focuses on specific areas of critical investigation, and on a continuous exercise seeking a reflective research hypothesis and an emerging methodology. The residencies at different laboratories were sustained under research methods familiar to my practice and an ethnographic approach has been used, with studio work mirroring an evolving embedded research process from fieldwork. The example PhD [Lopes, 2013] constructs connections between visual art practice and neuroscientific research studies in the field of Alzheimer's disease (AD). The aim was to explore the representational strategies of AD studies in the laboratory through art practice, alluding to the (dys)functioning of autobiographical memory. Debates on visuality and 'looking at', the institutional gaze and the technology within which the visual and science are rooted are examined using Foucault's concepts of heterotopia and panopticon and Latour's actor-network theory [Lopes, 2007]. In creative terms, this exploration allowed the development of several novel and unique methods of research, the resulting artefacts, and the practical materialization of these via the agency of installation. This is suggested as a fertile area for new art-based exploration and critiques of developments in both contemporary art and science. This presentation and article suggests that while exploring a space for art practice within the discourse of memory embedded in scientific laboratories and medical practices dealing with the network of people involved, I have gained a space within the discourse on arts and health.

5.1 Artistic residencies in scientific laboratories

My experience at artistic residencies in scientific fields began with a collaborative project (i.e. *Ethology I*[1]) developed at Instituto Gulbenkian Ciência (IGC) in Oeiras between 2006 and 2008. The project was supported by *Ectopia* an Experimental Art Laboratory for art/science projects that was being established and was innovative in Portugal. It was conceived as a collaborative transdisciplinary project involving an artist/director (Maria Manuela Lopes) and several scientists from different fields of study ranging from immunology and molecular biology to artificial intelligence [Lopes, 2007]. It made me realize that the collaborative model necessarily for approaching highly specialized fields of research was in fact communication between people. This experience made me deal with a new framework for my art creation, distinct funding policies, and institutional roles and ended up in a scale model, a computational prototype and a reflective experience presented at a few venues. This experience leveraged the proposal for the first institutional art/science residency program in Portugal, The Residences Network Programme: Art/Science/Technology Experimentation, a cooperation between Ciência Viva and Direcção-Geral das Artes (DGartes)/Ministry of Culture. I developed *Cage SSI* (Fig. 5.1), in collaboration with Alexandre do Carmo and CAGE research group at IBMC (Instituto de Biologia Molecular e Celular) in Oporto, which uses biotechnology and immunology to consider site-specific issues and symbiotic relations. The resulting interactive installation, with four superimposed video projections, is inspired by the mechanics and dynamics of the immune system, using bench work and immunological synapses as the foundation for video microscopy.

---

[1] *Site Specificity and the Interactive Space of the Human Body" and "Ethology I – Immune System,* ECTOPIA, Instituto Gulbenkian Ciência, in collaboration with Instituto de Medicina Molecular, INETI, Electronic Department Faculty of Science – Lisbon University, Antrophology Department - Coimbra University. Aknowledgments: Luís Graça, Marta de Menezes, Pedro Rocha, Paulo Urbano, Jorge Carneiro, Luís Quintais, António Jacinto, Jocelyne Demangio and Paulo Bernardino. [2006/2008]

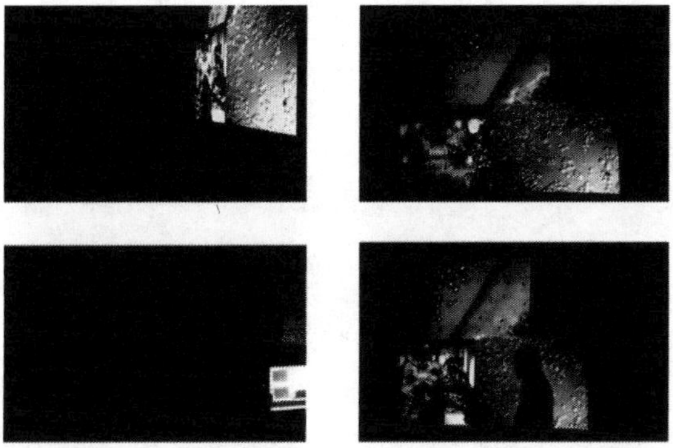

Fig.5.1 Maria Manuela Lopes, 2008, *CAGE SSI,* at *Exposição Experimentação: Arte|Ciência e Tecnologia,* Pavilhão do Conhecimento - Ciência Viva, Lisbon.

Later, in 2009, I went back to Ectopia and developed with Marta de Menezes, in collaboration with Isabel Gordo, *Tetrahynena*, in which we explore the concept of gender through a live organism that has seven distinct genders. The common aquatic organism Tetrahymena exists in Nature in seven different "mating types" (i.e. sexes). Individuals from each "mating type" display characteristic morphologic and behavioural traits and are able to mate with individuals from any other of the six types different from their own.

We use the live organisms as art media to generate an artwork addressing how each one of us relates to gender, sex, and mating preferences. In other words, we have chosen to develop a thought experiment, using objects and contraptions to challenge our preconceptions, belief systems, and rationality. Therefore, the artists' aim is not only to develop and experiment with biological material and techniques to develop and produce the artwork, but more importantly to experiment with concepts and preconceptions about gender issues that concern us all.

Tetrahymena is publicly presented as an installation with a number of pieces that include the live organisms, video, and reflections of the observers. These installation pieces sequentially question the diversity of the different genders, or mating types; our own position within such diversity; and the supra-individual relationships that form the basis of communities and societies.

The idea of pursuing further research in the themes of memory related to biological and medical studies lead me to my practice-based PhD in collaboration with Neuroscientists at the Hospital Santa Maria and Instituto de Medicina Molecular in Lisbon. The research project had a multidisciplinary supervisory team constituted by an artist, a neurosurgeon and a cultural theorist (namely Kathleen Rogers, João Lobo Antunes and Judith Williamson), and the context provided a unique and complementary scientific/laboratory/artist studio in which to develop a progressive and situated practice-based inquiry into how memory loss is scientifically researched.

5.2 Representational Strategies on Alzheimer's Studies: A Practice-Based Art Research in a Neuroscience Laboratory

Recent neuroscientific research, centred on the neurophysiological case studies of patients with brain injuries as well as cognitive studies on healthy subjects, suggests that memory is in fact a multiple memory system [Tulving, 1972; 1985]. Memory seems to be not a single entity, but a group of different abilities mediated by distinct brain systems, such as the sensory cortex, the dorsal striatal, the amygdala and the hippocampal region. Each neural system holds differing connections and receives different information from distinct brain structures. The working of these neural systems is understood as both independent and cooperative, which explains dissociations in people with brain lesions, as well as the unified experience of memory in a normal person. This view is supported by advanced functional neuroimaging investigations [Schacter et al., 2000].

In neuropsychology, when a test is performed on a patient's memory, the meaning associated with memory is, according to Tulving and Craik [2000], a neurocognitive capacity described as a three-stage process of encoding, storage and retrieval. The process of encoding the links to the initial absorption of information into the memory leaves traces in the nervous system that preserve the effects of experience across time. Storage is the maintenance of these traces. The act of remembering in neural terms is described as a network of connections among nerve cells [Kandel, 2006]. Information arriving from the world is stored in the limited-capacity process of short-term memory, before being lost or transferred to a separate long-term memory system.

Memory is currently seen as an activity under constant construction and reconstruction [Rosen, 1998], subverting the traditional view that it is a storage machine for record, retention and unaltered retrieval [Sutton, 2004]. Subjectively, it is understood that memories are mediated, personal and subject to change, rather than a fixed archive of pristine data awaiting the appropriate command to re-form them into images. This subjective weaving quality allows psychologists to qualify them as autobiographical memory.

Psychologist David Rubin [2005, p. 79] claims that many systems play a role in autobiographical memory, I propose that this process in a multiplicity of systems implies a tendency to assemble, and may be evoked through montage, juxtaposition and video installation. In Alzheimer's disease communication within this system seems to be disrupted; the assembling capacity therefore seems to be shattered.

Alzheimer's is a silent, neurodegenerative disease that effaces the map of connections drawn from the experience of living; its onset is heralded by destruction of autobiographical memory. According to neuroscientist André Dellacourte [2006], Alzheimer's disease is the most common form of dementia, a brain pathology that leads to a progressive cognitive decline in the ability to remember, learn, think and communicate. The progressive loss of brain network connections inhibits patients from assessing not just recent events, but well-rehearsed skills and known facts.

Science and health philosophers Marc Berg and Annemarie Mol [1998] claim that medical and scientific investigation is based on practices of construction, assemblage and incorporation of techniques, habits, ideas, images and conversations. This represents my concept of an archive, envisioned as a network of four intertwined 'archival spaces' (Fig. 5.2):

- The patient;
- The art studio;

- The Alzheimer's research laboratory;
- The cellular and molecular research laboratory.

Fig.5.2 Maria Manuela Lopes, 2009. 'The four virtual archival spaces.'

In this process I have documented the assessment and categorization of Alzheimer's patients and their exposure to various therapies. I have also documented molecular and cellular research laboratory procedures and collect laboratory materials. In order to maintain a common thread when approaching these spaces, methods have focused on neuroscientific diagnostic images (MRI and PET scan), graphs, diagrams, notes, models, photographs, instruments, physical space conditions and hybrid forms of these, and establishing a critical link to art practice. In laboratory studies these visual approaches are designated representational strategies or practices [Lynch and Woolgar, 1990, p. 1-18].

Analysis of my working methods identified the need for some studies on the status of images, visuality and representation, due to the interdependency of my ethnographic exploration on vision and technology. This reflection served to critically consider my voyeuristic position, using the artistic gaze on others' practices and diseases to question the roles of observing, archiving and reconstructing. For example, Jonathan Crary [1992] focuses on the embodied observer's ways of looking. He examines how practices were shaped by optical devices, including the camera obscura,[2] panoramas, magic lanterns, kaleidoscopes and stereoscopes, thus ascribing supposed neutrality to the technologies leading to codified modes of viewing.

Reflecting on the idea of 'looking at' in institutions, French philosopher Michel Foucault explores the ways in which institutions such as schools, prisons, hospitals or asylums depend on several forms of surveillance. In his 1967 lecture 'On Other Spaces'[3] he identified specific types of spaces that have links to all other sites in our culture, but in a way that contradicts or reflects the set of relations they designate. Places such as cemeteries, theatres and museums are termed 'heterotopias', and this metaphor has been used to describe my network of archival spaces because of a recognized similarity of principle.

Foucault views 'crisis heterotopias' as places reserved for individuals in a state of crisis with respect to society and the human milieu [1998, p. 179], though he refers to transitory crisis such as adolescence, citing examples such as educational experience or military service. This characteristic corresponds broadly to two of my archival spaces: Alzheimer's patients and the hospital/clinical research unit, because of their association with a state of crisis. Another principle refers to connections and relations simultaneously juxtaposed and previously non-existing, exemplified by the way in which theatre and cinema provide symbolic meanings. In my research these correspond to the studio, where montage and reconstructions are actuated and new connections forged.

An additional principle presupposes a system of opening and closing, isolating heterotopias and thus at the same time rendering them penetrable, either through coercion, as in imprisonment, or via rituals and

[2] The camera obscura is a common metaphor for memory. In *Metaphors of Memory: a History of Ideas About the Mind*, Douwe Draaisma [2001] presents an interesting case study of the way in which descriptions of memory, supported by the vague but continuous and dynamic images of the *camera obscura*, rapidly gave way in mid-19th-century psychology to the photography-inspired contradictory notion that memories are fixed and accurate but static.

[3] Lecture presented to the Architectural Studies Circle 14 March 1967 and published as 'Different Spaces' in Faubion, James D., ed. [1998], *Essential Works of Foucault: Aesthetics, Method, and Epistemology*, vol II. The Penguin Press, London.

purifications. I see a similarity with scientific laboratories, as places that are only accessible with special permission obtained by following bureaucratic and aseptic procedures. The processes of filling in various consents, contracts, health and safety or ethics forms, in addition to procedures such as donning white coats and cleansing of hands, are seen to equate to ritualistic and purification performances.

In *The Birth of the Clinic* Foucault argues that the mechanical-clinical gaze transforms our bodies in sites where organs and eyes meet [2003, p. xiv], detaching the body from the person, or 'externalizing the internal'. He explains that this gaze allows a passage from small-scale practice to an archival kind of functioning, substituting body presence for large manipulation of coded records permanently present. This approval [Struken and Cartwright, 2002, p. 281] of the instrumental image as a provider of access to invisible truths is linked to an increased use of institutional regulation and categorization or archiving. This allows the clinical gaze to extend from the individual body to the wider population, and a stronger weaving of technology, culture, medicine and society.

The late or scarce physical symptoms of Alzheimer's disease prompt my interest in how the intermediation of the technological image substitutes for some clinical sensorial examination of the body, and how certain laboratory practices turn the biological materiality of the loss of memory into a visible phenomenon. In clinical practice, perspectives similar to those adopted by Crary or Cartwright assume a supposed neutrality of the technologies and ascribe objectivity, methodological consistency and precision to the images.[4] Representation is the substitution of symptoms and behaviours reported by the patient. Images are seen as essential to systems of power and concepts of knowledge and self-regulation. Foucault uses a powerful metaphor to explain this idea: the panoptic[5]

---

[4] 'The notion of photographic truth hinges on the idea that the camera is an objective device for the capturing of reality, and that it renders this objectivity despite the subjective vision of the person using the camera.' [Sturken and Cartwright, 2002, p. 280].

[5] Foucault uses the design of a model prison conceived by reformist Jeremy Bentham (1748-1832). In Bentham's model: 'at the periphery, an annular building; at the centre, a tower; this tower is pierced with wide windows that open onto the inner side of the ring; the peripheric building is divided into cells, each of which extends the whole width of the building; they have two windows, one on the inside, corresponding to the windows of the tower; the other, on the outside, allows the light to cross the cell from one end to the

prison functioning as a reminder of the permanent exposure to visibility to which one is from then on subjected. As Foucault [1995] suggests, scientific and medical understanding of the body focuses on surveillance using practices of assessing, recording and archiving to classify and define the boundaries between illness and health.

An apparent organization of the sensorial perception of space and time thus articulates power relationships. According to Struken and Cartwright [2002] this looking is linked to other activities that ascribe meaning to what vision uncovers, such as experimenting, measuring, analysing and ordering. My practice uses a similar approach to deal with reorganisation of materials, space and time around the study of Alzheimer's. The panoptical gaze surveys the human body, which for the purposes of the clinical investigation becomes transparent; the invisible disease becomes accessible; volumes, spaces and time become visible in representational strategies, both portable and juxtaposable in a translation format. Vision prevails but is dependent on other sensory and cognitive processes, and also on context.

The most radical example of this practice was the development of the X-ray as a medical representational strategy. The opening of the clinical gaze to a wider audience is explored in several studies that focus on transparency and approach of the body's interior, an experience only possible via X-rays and medical imaging techniques such as ultrasound, MRI and PET scans.

The relationship between technology, science and nature is explored in Bruno Latour and Steve Woolgar's seminal study of 'laboratorial life', and Latour's studies around action in science.[6] Latour [1986] suggests

---

other. All that is needed, then, is to place a supervisor in a central tower and to shut up in each cell a madman, a patient, a condemned man, a worker or a schoolboy. By the effect of backlighting, one can observe from the tower, standing out precisely against the light, the small captive shadows in the cells of the periphery. They are like so many cages, so many small theatres, in which each actor is alone, perfectly individualized and constantly visible.' [Foucault, 1995, p. 200]

6 These are particularly useful and new insofar as they interweave discourses on biotechnology research and production of facts and knowledge using an ethnographical approach. The reader is driven through different construction sites of facts and machines staging science as an inscription-making practice, prompting discussion of issues around representation and inscription and science as a semiotic practice dealing with signs and symbols, presence and absence.

that science is a process of addition of inscriptions, which primarily works through study of images, graphs, maps, tables and text data rather than the real phenomena. He emphasizes the advantage lent by such images and inscriptions over scientific claims, with a focus on the whole process of mobilization surrounding their production. Latour's analysis suggests that observing means constructing and bringing into being. This applies particularly to technologies such as MRI and PET scans, in which observation by construction is mediated and functions at the very boundary of what can be conceived as visual.[7]

Bruno Latour, Michel Callon and John Law developed the actor-network perspective (or theory, i.e. ANT) on scientific knowledge, taking Foucault's power/knowledge understanding as a starting point but focusing on networks. Actors may be objects as well as people. The network suggestion allows entities to circulate in the laboratories and be exported as scientific facts or representations following processes of scaling down, inscription and construction. The conditions of art/science 'collaboration' in this project derive from my understanding and appropriation of metaphors from the ANT allowing a free circulation and non-hierarchical use of representational strategies from each 'archival space'. The word collaboration when used can be understood as facilitation. Networks are intended to connect laboratories to chains of actors in different spatial locations and to address the contradiction that so few people seem to embrace the world.

---

[7] For example, MRI depends on the use of very strong magnetic fields to make the nuclei of atoms spin in a certain direction. This effect is fundamentally undetectable in any form by the human senses, although the amount of noise might be misleading in this respect. In contrast to a microscope that magnifies visible light, the sophisticated measurement of electromagnetic energy, outside the scope of human sight, must be reconstructed in a pictographic way that may be interpreted by expert technicians and doctors. Accordingly, not only is the interpretation subjective, but also the visual choices made in the graphical interpretation of the scan data for diagnostic purposes may obscure other readings; norms and consensus are under constant discussion in terms of choices of how to represent the scaled measurement of the magnetically spinning nuclei in a visual way.

Latour [1987] emphasizes the central role played by images such as inscriptions and material artefacts in consolidating the networks established between laboratory and the outside community. He calls these delegates 'immutable mobiles', which should provide important information of the places and people concerned. The concept of 'translation' is used to a practical end: anyone who needs to construct a fact must involve other people and technologies, but must simultaneously control their performances. He calls attention to the interpretation these actors hold on relating their own interests in the construction of the facts. The term 'translation' is chosen for its linguistic meaning – relating versions in one language to versions in another – and for its associated geometric meaning of moving from one place to another [1987, p. 117].
If in the laboratory images are mainly read against the background assumptions of medical discourse and practice, in the studio they surface under other cultural assumptions. Through different visual strategies, my artistic work deploy metaphors relevant to the current neuroscientific approach to memory and biological research, and also explores a cultural critique that resonates within those theories.

In *Six Stories from the End of Representation*, James Elkins [2008, p. 18] proposes an exercise to build bridges between the humanities and the sciences. This involves allowing them to stay side by side telling stories in their own languages, thus forcing the viewer to decide which threads may tie the presented images together. In my work, fragments of text, images, sound and time are enacted and re-enacted into art practice, during which the scientific materials are re-framed within cultural and social conventions. This process locates them in the history of artists such as Helen Chadwick, Suzanne Anker and Mark Dion, who have explored scientific laboratories and medical imaging and practices in the form of personal portraiture, commentary on biological identity, nature, the interweaving of culture and technology, as well as critique on practices of looking, ordering and archiving. My engagement with the scientific research renders the process of archiving contingent on my subjective experience. This contextual immersion depends on conceptualization and reflection of my artistic role in their scientific ground. My work therefore involves challenging scientists in this specific field of neuroscience to adopt a visual approach to their research topic.

5.3 Conclusion

My art practice and research in scientific laboratories has explored the dynamics of the encounter between diseased bodies, institutionalised scientific practices and diagnostic images, focusing on the enactment and re-enactment of the originating performances, their experiential sensory qualities and their power to address the subject of memory and identity. These findings have been extended into cultural, social and political contexts by contrasting them with other 'realities' and private/public cultural practices.

The recent PhD project has shown that our fragmented memories, body and relationship to objects, language and knowledge are characterised by paradox, fluidity and challenge in a permanent search for meaning. As a result, it is possible to address the notion of wholeness via an installation (Fig. 5.3) that is presented to the viewer as a new network of potential narratives about autobiographical memory, language, knowledge and self. These elements are connected within neuroscientific studies of Alzheimer's disease, the patient, and my art practice, but their relationship cannot be comprehensively defined. In contemporary understanding, memory is a constant reconstructive subjective process, as are the scenarios depicted in the installations. These may be summarised as the re-privileging of the aesthetic experience of memory, framed within the matrices of both an innovative art/science collaborative model and the evolving process of practice-based research in the visual arts, in order to generate fluid, hybrid outcomes.

Fig. 5.3 Maria Manuela Lopes, 2013, *TRACE,* at *KN SINGH | Studio Farnham, UK*

The representational strategies I have used in my research have not been limited to the visual, but included anything that could be viewed as a conduit for a message including images: elements of language; objects; protocols; location and moments in time; its premise is that art practices can include the use and manipulation of scientific practices, codes and images to produce effects and affects.

Within and across the scientific field, images, gestures and objects require labels and instructions for use, in the form of explanatory or contextual text. These are usually archived and transferred with related images or other visual representations, fundamental epistemic elements in the scientific network. My residencies in scientific laboratories and the collaborations have shown that when deprived of their associations, however, inscriptions become iconised or recontextualised in novel ways.

On an epistemological level, the bridging and reciprocity between science and art and interdisciplinary practice-based research projects may contribute to the introduction of a more open research paradigm in contemporary research. I conclude that while exploring a space for art practice within the discourse of memory embedded in scientific laboratories and medical practices dealing with the network of people involved, I have gained a space within the discourse on arts and health.

References

Berg, M. & Mol, A., eds. [1998] *Differences in Medicine: Unravelling Practices, Techniques and Bodies*. (Duke University Press, USA).

Crary, Jonathan [1992] *Techniques of the Observer: On Vision and Modernity in the Nineteenth Century.* (MIT Press, Cambridge, MA).

Dellacourte, A. [2006], *Alzheimer's Disease definition*, at: http://www.alzheimer-adna.com/Gb/Alz/Alzheimer_clinical.html (accessed 07.02.2009).

Draaisma, D. [2001] *Metaphors of Memory: a History of Ideas About the Mind.* (Cambridge University Press, Cambridge).

Elkins, J. [2008] *Six Stories from the End of Representation.* (Stanford University Press, Stanford).

Foucault, M. [1995] *Discipline and Punish: The Birth of the Prison*. (Vintage, New York).

Foucault, M. [1998] *Different Spaces*, in Faubion, James D., ed., *Essential Works of Foucault: Aesthetics, Method, and Epistemology*, vol. II. (The Penguin Press, London).

Foucault, M. [2003] *The Birth of the Clinic: Archaeology of Medical Perception*, Taylor and Francis e-Library.

Kandel, E. R. [2006] *In Search of Memory: The Emergence of a New Science of Mind.* (WW Norton, New York).

Latour, B. [1987] *Science in Action: How to Follow Scientists and Engineers through Society.* (Harvard University Press, Cambridge, MA).

Latour, B. & Woolgar, S. [1986] *Laboratory Life: the Construction of Scientific Facts*. (Princeton University Press, New Jersey).

Lopes, M. M. [2007] *Site Specificity and the Interactive Space of the Human Body and Ethology I – Immune System*, (TECHNARTE - International Conference on Art & Technology – Bilbao, Spain).

Lopes, M. M. [2011] *Re-enacting the Self in the Archive: Memory, Art, and Neuroscience,* (Proceedings ISEA –International Symposium on Electronic Art, Istanbul, Turkey).

Lopes, M. M. [2013] *Representational Strategies On Alzheimer's Studies: A Practice-Based Art Research In A Neuroscience Laboratory*, (PhD dissertation, unpublished, University for the Creative Arts – Farnham, UK).

Lynch, M. & Woolgar, S. [1990] "Introduction: Sociological Orientations to Representational Practice in Science", in *Representation in Scientific Practice,* eds. Lynch, M and Woolgar, S. (MIT Press, Cambridge, MA) pp. 1–18.

Rosen, H. [1998] *Speaking from Memory: The Study of Autobiographical Discourse.* (Trentham Books, Staffordshire).

Rubin, D. C. [2005], *A Basic-Systems Approach to Autobiographical Memory*, at http://search.yahoo.com/r/_ylt=A0oG7m6u5PNQlD4ARLxXNyoA;_ylu=X3oDMTE1M2szdm04BHNlYwNzcgRwb3MDMQRjb2xvA2FjMgR2dGlkA1

ZJUDE5NV8yMDk/SIG=137grplch/EXP=1358189870/**http%3a//www.brown.edu/Departments/Humanities_Center/events/documents/cl175Rubin2.pdf (accessed 22.11.2009).

Schacter, D. L., Wagner, A. & Buckner, R. [2000] "Memory Systems of 1999", in *The Oxford Handbook of Memory*, eds. Tulving, E. & Craik, F. (Oxford University Press, Oxford) pp. 627–645.

Sturken, M. & Cartwright, L. [2002] *Practices of Looking: an Introduction to Visual Culture* (Second Edition - Oxford University Press, Oxford).

Sutton, John [2004] 'Memory', *Stanford Encyclopedia of Philosophy*. http://plato.stanford.edu/entries/memory/ (accessed 10.10.2009).

Tulving, E. [1985] "How Many Memory Systems are There?", *American Psychologist* 40(4), 385–398. http://alicekim.ca/14.AmPsy85.pdf (accessed 16.01.2010).

Tulving, E. & Craik, F. [2000] *The Oxford Handbook of Memory.* (Oxford University Press, Oxford).

Tulving, E. [1972] "Episodic and Semantic Memory", in *Organization Of Memory,* eds. Tulving, E. & Donaldson, W. (Academic Press, New York) pp. 381–403.

6

Exploratory Augmented Reality Art Based Projects in Anatomic Pathology Laboratories: RAMPAAP

Sérgio Eliseu, Paulo B. Bastos, Maria M. Lopes and Manuela Monteiro

This article describes an art work in progress (RAMPAAP)[1] which combines Augmented Reality (AR) techniques with the methodologies exercised in two specific Anatomic Pathology laboratories (one researching cancer and the other dementias)[2]. It also deals with the research results obtained by members of our team in previous successful scientific collaborations, which were revealed in numerous publications and public exhibitions. As artists and researchers, we are devoted to explore possibilities to extend the knowledge through visual and material practices and empirical methods. So we propose to investigate how AR and more immersive display technologies can be explored in order to produce high quality science informed artistic projects and also to generate interfaces that can be used within the realms of medical education and training. The areas included in this project, artistic, health care and technology of AR, are original research, innovative and vital to boost the Portuguese artistic, scientific and technological scene, as well as medical education and public awareness in general. We believe that is and will continue to be a challenge in the near future, considering the increasing use of AR and the growing percentage of elderly population along with the dramatic number of patients with dementia and cancer worldwide, as well as the overwhelming predictions of its growth.

---

[1] R.A.M.P.A.A.P. is an acronym that stands for the project Portuguese title "Realidade Aumentada versus Realidade Melhorada: Projeto Artístico (exploratório) em Laboratório de Anatomia Patológica"

[2] Serviço de Anatomia Patológica, Instituto Português de Oncologia de Francisco Gentil (IPO Coimbra) and Newcastle Brain Tissue Resource, Institute of Ageing and Health, Newcastle University.

6.1 Introduction

Artists and doctors have shared their interests for a long time. Indeed, there are many examples where, essentially at the level of representation of body anatomy, the research of both has crossed throughout the past. Numerous illustrations have been created by artists hired by doctors and/or doctors trained as artists. Albrecht Dürer, Leonardo Da Vinci and Pietro Berrettini, for example, studied anatomy, although outside the medical context. The Scottish brothers John Bell and Charles Bell were trained as artists before they became surgeons. The French anatomist Jules Cloquet, in the XIXth Century, studied art before becoming a physician and William Hunter, anatomist and obstetrician, was also dedicated to the teaching of anatomy for artists [Adelino et al, 2010].

Despite the fact that many examples could be found where science, medicine and art encounter each other, either centred in only one person or in group initiatives (such as the ones founded by Wellcome Trust or Arts Catalyst in the UK, or Ectopia, Cultivamos Cultura with DGArtes or Ciência Viva in Portugal), in everyday practice art and science still look at events from distinct perspectives. Certainly because there is a high degree of expertise and a great complexity of the techniques used. However, this fact does not necessarily dilute, but it may even enhance the enduring relationship between artistic production and medicine. No longer located only in the field of two-dimensional representation, but in multifaceted projects, ranging from the macabre installations of Gunther von Hagens to the relationships that artist Íñigo Bilbao experiences with biomedical imaging and many others that already embrace interactivity [Eliseu et al, 2012]. Thereby, many studies are now focused on visual strategies common to science, medicine and art, questioning the externalization of the invisible body and the relationship between seeing and knowing. Some other studies even explore art-science collaborations close to technological and social advances in molecular biology reflecting on the use of those materials, processes and theory by artists. Those projects are attempts to decode the cultural codes also enclosed in the scientific images in a bid to make them a space where other fields of culture may be inscribed. They do this by questioning their claim of 'objectivity' and 'truth' and opening them to other areas of knowledge. [Reichle, 2003].

Embedded in this paradigm, RAMPAAP was born from the crossing of a collaboration between a researcher and visual artist of our team and the Department of Pathology of the Portuguese Institute of Oncology in

Coimbra[3], and from a doctoral arts research project of another member of the team[4] that included a period of residence in two distinct groups of Santa Maria Hospital and Institute of Molecular Medicine to investigate the dementias by several pathways (cellular and clinic)[5]. The art projects developed and publicly presented in both cases lead us to conclude that there is a very rich plasticity and an expressive metaphorical meaning in the several materials and practices used in a laboratory of anatomic pathology that deserve a further exploration. And that, on the other hand, the applicability of the technology of AR to new laboratory techniques is very promising, mainly in the areas of artistic communication, awareness and teaching. Thereby, in RAMPAAP, our main goal is to contribute to the enhancement of reality in the research across the above mentioned fields, adding a surplus socially engaged value for addressing an ethically challenging theme such as cancer and dementia pathologies.

6.2 Introducing Augmented Reality (AR)

AR[6] technology corresponds to the overlap of virtual objects to the physical environment, which can be viewed, in real time, through technological devices [Kirner, 2004]. Even though it was long before it was even technically possible, the idea of AR appeared in "The Master Key", a book published in 1901. Written by L. Frank Baum (1856-1919), this literary work describes a pair of spectacles which are able to add information to reality. However, the first use of AR, still without this designation, dates from the 1950s, when Morton Heilig, a film director, developed a stationary system (*Sensorama*) which encompassed all the senses of the spectator, providing a greater immersion and involvement with the film. *Sensorama* was capable of displaying stereoscopic images

---

[3] That explored and developed interfaces, as well as specific computer applications to allow an alternative display of interactive three-dimensional images in real time from various tissues in successive processing steps and diagnosis practiced in the laboratory (Sérgio Eliseu, since 2011).

[4] In this other project, Maria Manuela Lopes produced a practice-based PhD research and developed a body of four interrelated installation artworks.

[5] This research included following studies on the neuropathology service observing both morphological and macroscopic processes on human brains including microscopy and histology of nervous system tissues.

[6] The term "Augmented Reality" dates back to 1990 and is attributed to Tom Claudell [Mullen, 2011].

along with stereo sound, body tilting, wind and scent sensations [Carmigniani et al, 2011]. Nevertheless, the first technical study registered within the concept of AR dates from 1968 [Sutherland, 1968], when Sutherland developed a head mounted display (HMD) with three dimensional graphics (a cube) which were simultaneously presented with real objects.

This technology is considered as a subfield of virtual reality [Milgram et al, 1994], with the difference that virtual reality carries its user to the virtual environment and the user of AR is kept in his physical environment. That is, by using this technique, the virtual objects (which may be or include graphics, sounds or other sensory stimulation) became part of the real environment, in which they can be handled, involving the user in a more immersive, compelling and attractive interaction.

Fig. 6.1. "Ser Real", Augmented Reality art installation. Sérgio Eliseu. Aveiro Museum, 2009.

Nonetheless, the open up experiments in AR projects, which were performed until the end of the XX century, required the use of computers with high processing capacity, along with the necessary programming skills. As such, the research costs were highly enlarged, making it much more difficult to expand this technology and resulting mostly in stationary systems developed in laboratory settings. This situation was only overturned at the beginning of the XXI century, with the emergence of mobile devices (smartphones and tablets) with similar processing ability and multiple sensors. In fact, nowadays, the AR systems are associated with several forms of display. It is possible to develop AR in a fixed form (using the image processing in televisions, monitors, projectors, etc.) or mobile form (using smartphones/tablets, virtual reality goggles, etc.) where the virtual image is overlapped on the actual image captured by a camera or by means of projection systems that project a virtual image directly on a surface or in a transparent screen (that's the case of the recent eyewear Google[7] and Atheer Labs[8] research).

All this new possibilities allowed virtual elements to be generated in real time, within any real environment, as well as inexpensively. When coupled with the development of freely distributed software, this evolvement has been the main contributor to the recent popularization phenomenon regarding AR. As such, given the potential of the AR technology and the present ease for the development of this type of environments with free tools, its application in the arts is not only an exception, but also an opportunity for its fervent exploration and development of new concepts.

6.3 AR Assertion: achievements in the course of Art

Throughout history, artists have always searched for illusive and immersive representation forms trying, to some extent, to use the most advanced practices and inheriting, at each step of this research, not only the techniques, but also the reflections of previous models [Grau, 2003]. However, new technologies aren't usually developed with artistic purposes but by science and industry for social, more pragmatic and utilitarian ends [Pearson, 1988]. Thus, it's important to transcend the initial technological fascination. It is up to artists to act as pioneers,

---

[7] Google Glass is a device that provides a small screen above the field of view. It offers maps, music options, weather, make video calls or take photos/film and share instantly via the Internet. More info at https://plus.google.com/+GoogleGlass/posts (accessed 10.02.2013).

[8] http://www.atheerlabs.com/ (accessed 10.08.2013).

seeking new aesthetics that result from the implementation of these new technologies [Papagiannis, 2011]. In fact, since the late 60's and booming in the 70's, artists like John Cage, Nam June Paike, Ronald Nameth, Charlotte Davis, Brenda Laurel, among others, have acted as forerunners in the application of new media in the development of their work. Starting to discuss and evaluate the use of new technologies in the arts in a very early stage, these artists took this new kind of production well beyond the mere use of technological innovations. This attitude was also heavily mimicked and developed in the later works of artists like Neil Brown, Dennis Del Favero, Matthew McGinity, Jeffrey Shaw, Peter Weibel, among others, where we are able to encounter the development of some projects applying AR. Nevertheless, we may assert that the artistic exploration of AR began in the mid-70s, when Myron Krueger, a multimedia artist very influenced by the work of John Cage about indeterminacy and audience participation, led public interaction to new frontiers, becoming a pioneer in the human-computer interaction field. In a very early stage, Krueger created artworks that responded to the movement and gestures of the public through an elaborate system of sensitive floors, graphic tables and video cameras. In effect, this artist has produced some of the first digital interactive immersive environments of the art world, which may currently be classified as AR.

Fig. 6.2. "Videoplace", interactive installation. Myron Krueger, 1974.[9]

[9] www.inventinginteractive.com/2010/03/22/myron-krueger/ (accessed 10.08. 2013).

As a good example of the aforementioned, one could remark “Videoplace" (1974), an interactive piece where a computer responded to a subject’s gestures, interpreting and even anticipating his actions. People could "touch" their computer generated silhouettes, as well as handle some animated graphics that appeared on the screen. However, only many years later, in 1994, Julie Martin, wife and collaborator of Billy Klüver, produced what is nowadays considered as the first art project in AR [Mullen 2011] - “Dancing in the cyberspace”, a play where dancers interacted with virtual contents projected on stage.

With its unique characteristics, AR meets many of the needs of the exhibition and entertainment industries, among which are the following advantages: improved sense of immersion; maintenance of the real world; personal and appealing experience; absence of the need to create complex virtual surroundings [Huang et al. 2011].

6.4 Previous Anatomic Pathology Laboratories Collaborations

6.4.1 Haptic ARnatomy

Anatomic pathology has always had a very important role in medical development. It was with this that the knowledge of the human body took place that provided the correlation between anatomy and pathology and that the morphologic, histologic and cytologic description was held. It’s a branch of the medical sciences field and it is directed towards the study of structural and functional changes of organs, tissues and cells. It is mainly based in a macroscopic study followed by a microscopic, chemical, immunologic and molecular study of biological samples (organs or tissues, including postmortem examinations and the study of biopsy specimens). This method is primordially developed in a laboratory environment, giving its contribution to prevention, screening, diagnosis and prognosis of illness. The anatomical samples are received in the pathology laboratory where they are given a gross examination. In that examination they are dissected and samples that represent lesions and suspect areas, seen with the naked eye, are taken. Those samples are processed (dehydrated, diaphanized and embedded in wax). That process allows making a wax block, which allows the cutting of the sample and contributes to its preservation. These are then cut in the microtome gaining, this way, serial cuts at least three micrometres thick. These cuts are then stained and observed on the microscope. Hematoxylin and eosin are used routinely, which gives a general overview of the morphology, but several other specific histochemical methods exist in order to emphasize the arrangement of tissue morphology components and certain substances that can be present in the sample. Several other methods can be used to distinguish, characterize and determine states of illness according to the suspicions and concerns of the pathologists: cytopathology (examination of unleashed cells), immunohistochemistry (use of antibodies to detect specific proteins), in situ hybridization (identification of specific DNA and RNA molecules), cytogenetics (chromosomes visualization and analysis), flow cytometry techniques (cells immunophenotype) and electron microscopy (the use of a greater magnification in order to examine cell components).

The organic human matter and tissues observed in the pathology laboratory are fragile and simultaneously unique and precious; further they are prepared through the histological practices with chemical solutions sometimes highly polluting and health hazardous, demanding sophisticated laboratorial handle and care. Therefore there is a difficulty in manoeuvring the teaching, the scientific share of knowledge and public sensitization on these areas. This has been only overcome with imagery and descriptive actions on the real practice, sometimes informed (i.e. medical students) by visits to the laboratory [Eliseu et al, 2012].

With the above issue in mind, the project developed in the Department of Pathology of the Portuguese Institute of Oncology in Coimbra (Haptic ARnatomy) began with the following objectives: to explore and produce new forms and systems for three-dimensional visualization of tissues; investigate and develop innovative practices of interaction with the image combining studies of Art, Design and Biomedicine; create artefacts that can be used by the general public, sensitizing it to certain methods of work in the Department of Anatomic Pathology; develop art and design projects around the explored theme presenting them in conferences and cultural venues .

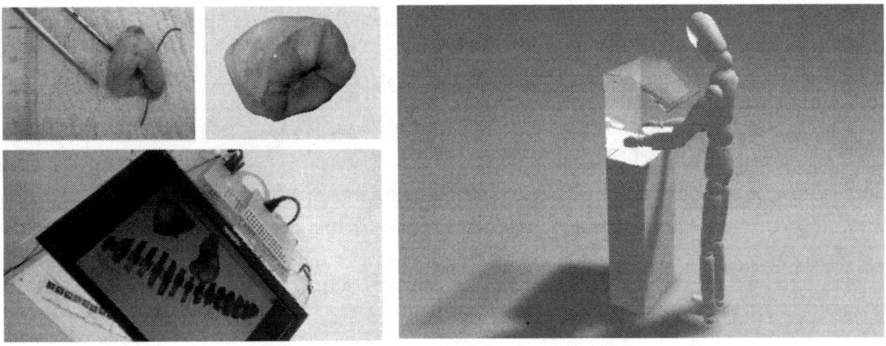

Fig. 6. 3. Prototype: "Haptic ARnatomy", Sérgio Eliseu, Pedro Cardoso. Fine Arts School, Porto University, 2012.[10]

As a starting point, we had tissue samples belonging to a lesion in the cervix. An anatomical piece was fully scanned at various angles and subsequently sectioned into 15 parts. All of them scanned after the appropriate and previously described processing steps. Next, three-dimensional models were produced manually using the editing software 3D Blender[11] through a textured bitmap (a bitmap or raster relates to an image composed of a two-dimensional array of points, in this case a regular grid pixel) obtained on scans. This was a lengthy process with a high degree of approximation, but unfortunately without achieving the perfect direct correlation that we wished in the beginning of the project.
The design of the artefact has been considered mainly in its relation with the audience, defining their possible settings and their tangible ergonomic user interface. Moreover, the software has been developed using a free computer vision library (ARToolkit) written and developed by Hirokazu Kato[12] that is extensively explored for applications in AR.
In practice, the tangible interface resorted to fiducial markers: imagistic patterns visually recognized by the computer system that serves as a reference point for spatial location and scale of the respective images shown in the project. The 32 markers corresponded to the number of options and sections made in anatomical piece. The user could freely use

[10] In the first prototype three dimensional models were produced manually and textured with bitmap images scanned from real tissues.
[11] www.blender.org (accessed 15.05.2015)
[12] Kirner [2008]

the parts where the respective markers, three-dimensional viewing each section and / or a record of each microscopy.

We presented two versions of the artefact. The first presentation occurred in October 2011 at the Museum of Fine Arts Faculty of the University of Porto and was part of the exhibition "Orality, future of art?" included in the Art and Design PhD students course. The purpose of the installation was to propose the question of the value of the object/image out of context institutional scientific, favouring on an enjoyment/closer interaction and subjective aesthetic merit. Away from the laboratory context the artefact gave rise to new readings, this time with cultural assumptions rather than the structure of the original scientific analysis. From a text of Ernesto de Sousa[13], "Orality, future of art?" the maximum "important is to get started" served as a catalyst to the research process of thinking and doing. Thus, leveraging new and imaginary narratives, the experience/enjoyment of the artefact broke the stereotyped dichotomy of creator/viewer. Raising a global involvement and avoiding setting a passive look, the images acquired new meanings exceeding their material basis.

[13] José Ernesto de Sousa (1921, 1988). Founder of the film society movement in Portugal, he was an artist, researcher and art critic.

Fig. 6. 4. Prototype II: "Haptic ARnatomy", Sérgio Eliseu, Pedro Cardoso. CAE - Center for Performing Arts - Figueira da Foz – at the XIII Congress of Pathology, 2012.

The second version prototype presentation took place at CAE - Center for Performing Arts - Figueira da Foz – at the XIII Congress of Pathology|2012. Here the artefact was presented with slight morphological changes in relation to the first display and the interaction of the parts turned to glass slides with original fiducial markers on both sides. The device was installed along with other pieces in a museum space, oriented to a scientific structured point of view of the several stages in the evolution of strategies for macro and microscopic analysis. There, the interpretation achieved an epistemological value, arousing curiosity about its practical viability in future implementations. In fact, the developed concept found a fertile ground in the laboratory settings and the applicability of a new perspective was established under possibility of a direct correlation between the three-dimensional models and the actual tissues. Consequently, as result of this Art-Science collaboration we expect to develop very soon a user friendly mobile version and continue to extend the perspectives and boundaries on science and art practices as research.

6.4.2 The four virtual archival spaces

In the other project, Maria Manuela Lopes produced practice-based PhD research and developed a body of four interrelated installation artworks (see chapter 5, this volume). The research constructs connections between visual art practice and neuroscientific studies in the field of Alzheimer's Disease (AD). The aim was to explore representational strategies of Alzheimer's studies in the laboratory through art practice, alluding to the (dys)functioning of autobiographical memory.

Fig. 6.5. 'The four virtual archival spaces', Maria Manuela Lopes, 2009.

Key elements include the visual, ordering, archiving and montage. The research acts as a metaphorical toolkit for providing a wider understanding of the practice of montage, archiving and time in connection with specific practices of neuroscientific research. Using successive artwork installations at different venues to elaborate on several simultaneous methods, the work reflects on how images and sounds relating to AD research may be understood as interacting elements which, when enacted and re-enacted, produce certain effects and affects that aim to account for the nature of memory and the disease itself.

Theories of autobiographical memory and current neuroscientific research on AD (clinical neurology, neuropsychology, neuropathology, neuroimaging and molecular and cellular studies) are introduced and explored. Debates on visuality and 'looking at', the institutional gaze and the technology within which the visual and science are rooted are examined using Foucault's concepts of heterotopia and panopticon and Latour's actor-network theory.

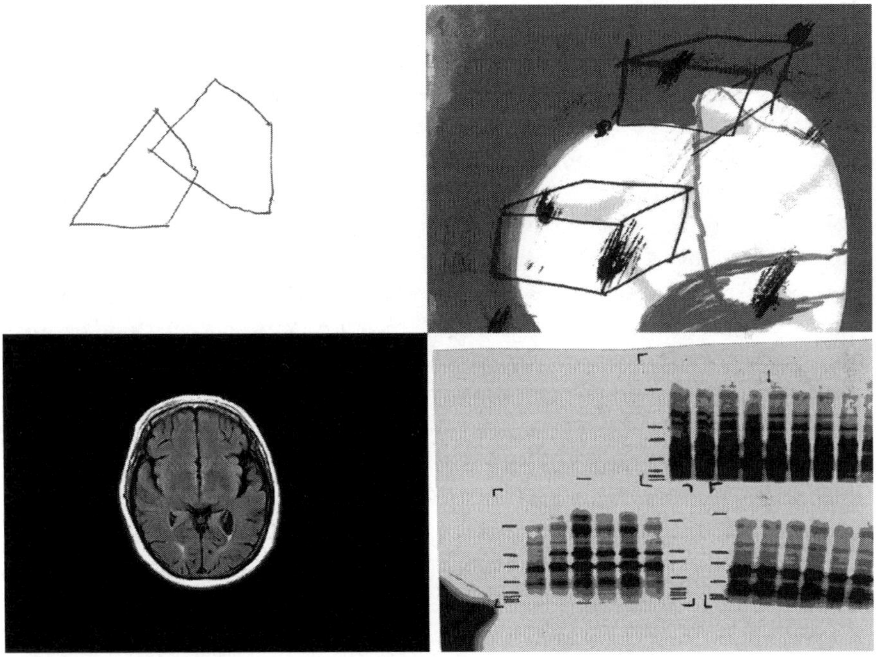

Fig. 6.6. 'The four virtual archival spaces', Maria Manuela Lopes, 2009.

In creative terms, the major conclusions are the development of several novel methods of research, the resulting artefacts, and the practical materialization of these via the agency of installation. Overall they are infused with qualities such as the temporal, the invisible, the archive and the creative re-enacting character of biological cognitive processes and scientific representational strategies. This is suggested as a fertile area for new art-based explorations and critiques of developments in both contemporary art and science.

The resources where from anonymous specimens of brain and nervous system tissue used by the craft routine and discipline performed by scientists in the daily life of research into neurodegenerative disease. But the representational strategies used in this research have not been limited to the visual, but included anything that could be viewed as a conduit for a message including images: elements of language; objects; protocols; location and moments in time. It explores the dynamics of the encounter between diseased bodies, scientific practices and diagnostic images.

The project has been conveyed through a range of case studies and own praxis. Its premise was that art practices can include the use and manipulation of scientific practices, codes and images to produce effects and affects. It engages discursively with some of the multiple factors that influence both the neuroscientific study of Alzheimer's disease and artistic production exploring the dynamics of the encounter between diseased bodies, institutionalised scientific practices and diagnostic images, focusing on the enactment and re-enactment of the originating performances, their experiential sensory qualities and their power to address the subject of memory. These findings have been extended into cultural, social and political contexts by contrasting them with other 'realities' and private/public cultural practices.

Since the project is discursively located within the paradigms of autobiographical memory, neuroscience and the visual, it allows the boundaries between the philosophic and scientific disciplines of epistemology, ethics and aesthetics, clinical neurology, molecular biology and technology to become central in a relational network associated with remembering and forgetting, presence/absence, and concepts of identity.

Within and across the scientific field, images, gestures and objects require labels and instructions for use, in the form of explanatory or contextual text. These are usually archived and transferred with related images or other visual representations, fundamental epistemic elements in the scientific network. In contrast, as demonstrated and also revealed in the works of Anker, Dion and Chadwick, the circulation of images from science into other spheres follows the rules and guidelines specific to those spheres. When deprived of their associations, however, inscriptions become iconised or recontextualised in novel ways. If not with an explicit revelation of the kind for which scientists strive, at least with a willingness to contemplate new forms of knowledge.

6.5 R.A.M.P.A.A.P.

Today real experiences in the scientific practices are possible to substitute by representational practices or inscriptions. As the sociologist of science Bruno Latour [1990] suggests, science is a process of addition of inscriptions, which primarily works through study of images, graphs, maps, tables and text data rather than the real phenomena. He emphasizes the advantage lent by such images and inscriptions over scientific claims, with a focus on the whole process of mobilization surrounding their production. The invisible truth may be sustained by showing an image, suggesting that observing means constructing and bringing into being. This applies particularly to technologies such as MRI (Magnetic Resonance Imaging) or AR and kinetic technologies in which observation by construction is mediated and functions at the very boundary of what can be conceived as visual [Lopes, M., 2013]. Already in "The Birth of the Clinic" Foucault argues that the mechanicalclinical gaze transforms our bodies in sites where organs and eyes meet detaching the body from the person, or 'externalizing the internal' [Foucault 2003]. He explains that this gaze allows a passage from small-scale practice to an archival kind of functioning, substituting body presence for large manipulation of coded records permanently present. The approval of the instrumental image as a provider of access to invisible truths is linked to an increased use of institutional regulation and categorization or archiving [Foucault, 2003]. This allows the clinical gaze to extend from the individual body to the wider population, and a stronger weaving of technology, culture, medicine and society. As Foucault suggests, scientific and medical understanding of the body focuses on surveillance using practices of assessing, recording and archiving to classify and define the boundaries between illness and health [Lopes, M., 2013].

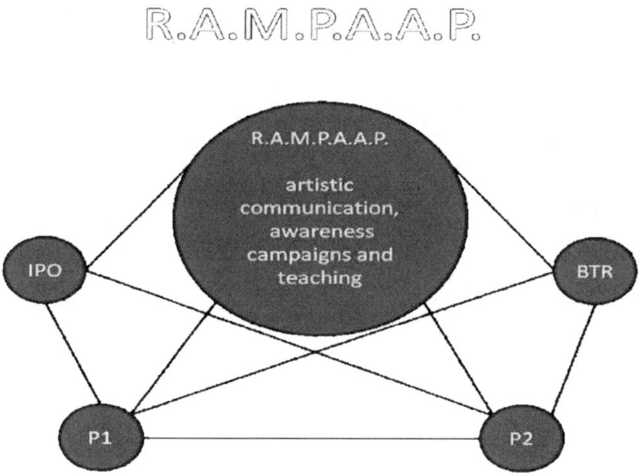

Fig. 6.7. "R.A.M.P.A.A.P." structure.

As artists and researchers devoted to explore possibilities to extend the knowledge through visual and material practices and empirical methods we propose to partially overcome the handicap that anatomic pathology education faces, by introducing AR visualization practices in the equation. We claim that we have a role to play when we consider not the end product (such as MRI, endoscopy or microscopy) but by exploring the underlying processes of constructing the images, in which the visual has a key function. Alongside this regard, our project has the following aims:

1- Departing from the existing interface prototype, to develop an AR interface and a computer based digital application for the visualization and interaction with pathology images of tissues and lesions;

2- To develop strategies for evoking the nature of the relationship body/health/identity/memory within a contemporary art practice, by generating art objects/installations using a period of residency at IPOCFG and in collaboration with the Institute of Ageing and Health;

3- To explore and construct connections between visual art practice and scientific research in the field of pathology;

4- To explore the representational strategies of cancer and dementia in the scientific laboratory through art practice;

5- To explore innovative and original experiences that include AR in the communication of pathology scientific practices and to present the artistic results in different venues, conferences and seminars, to differing audiences.

Previous experience in the laboratory and the development and testing of an AR interface led us to conclude on the need for a further exploratory period shared with distinct artists with novel approaches. Therefore the associations we were able to establish and the timeframe and budget limitation of the research leads us to propose a combined methodology. The methods in this exploratory research are multimedial, semi ethnographic and boundary transgressing and have been structured around four key stages: laboratory, art practice and design, application development, presentation and feedback. Some materials such as scientific inscriptions (e.g. microscopic images) are both methodological strategies and outputs. Linking this mobilization of multidimensional materials is part of the challenge of organizing the production of the artistic artefacts and of the AR interface, providing increased interdisciplinary activity. This interdisciplinary approach proves to be an asset, because the technical experts put at the disposal of artists and designers diverse technical capabilities that they would not be possible to grasp in the totality, and that is reciprocated with a reflectivecreative capacity that may enhance new approaches and use of technical knowledge [Eliseu, 2009].

Therefore this project in is exploratory and innovative nature aims at fostering an experiment targeted to produce artefacts; develop an interface for educational purposes; establish connections between artists and scientists towards a network of several institutions (hospital, laboratory, academia, artistic centres and distinctive audiences) and also intends to stimulate new views and theorizing within the humanities. The methods in this exploratory research are multimedial, semi ethnographic and boundary transgressing and have been structured around four key stages: laboratory, art practice and design, application development, and presentation and feedback.

Stage 1, Laboratory: this stage will take the form of a study of related works and the experience of artists (including our own) sharing the laboratory with the IPO pathologists and technicians in a quasi-ethnographic approach, (i.e. minimally invading the scientific practices). This stage will include the residencies of three selected artists interested in the framework of our research and in exploring AR technologies within the pathology laboratory of the IPOCFG, E.P.E. Experience in the clinical and scientific spaces will be oriented to observation, collection and video and photographic captures of repeated gestures, sentences, materials produced, and archival strategies. The findings, conclusions, recommendations and challenges that emerged from our previous experiences and that will resurface under the auspices of new inclusion of artists in this collaboration between visual artists, scientists and scholars will be used to refine our set of research objectives and questions and to design the exhibition needs.

Stage 2, Art Practice and Design: This stage will involve the creation of art projects and the design of the AR interface for education in pathology model, which describes the interplays and interrelationships between AR technology, education and medicine. We aim to produce some ebooks as diaries of the artistic explorations and publish them on the project site.

Stage 3, application development: Exploring, observing and testing will be the key words of this stage. This stage involves the developing of 3d models from the scientific images collected by the experts in pathology and pathology imaging production from IPO and IAH. These scientific end products are treated as the raw material for collaboration and will be critically addressed when transformed in an educational tool and interface that has to serve both purposes: to be as functional as the prototype was and to raise issues concerning the visualizing underlying processes, to appraise the wonders of the subtle, crafty revelatory processes of pathology.

Stage 4, Presentation and feedback: At this stage the team hopes to contribute to raise questions in a public venue where the AR interface for pathology education may be tested in a performative exploration side by side with three artefacts (possible AR installations) produced by the three artists in residence and also the research team, addressing managing the artistic process as a research endeavour.

6.6 Conclusion

Differences in culture are overcome by common interests. If we democratize and understand the pathologies and medical issues by exploring the underlying process of medical image production and human/machine interaction, we hope to enhance reality (practices) by also exploring creatively the technology of AR.

This team has previously successful experiences when working in scientific collaborations and laboratories that allowed us to claim that when crossing these territories the result is thrilling and that new transdisciplinary knowledge may arise in innovative understandings. On the other hand, we have encouraging evidence that scientific researchers are also attempting to open up dialogues and collaborate with artists in order to build on and share knowledge and cast a new light on their own work.

Thereby, with the goal of achieving interfaces and software applications of the interaction with real pathological tissues, and by extending the perspectives and boundaries on science and art practice as research, we also expect to achieve artistic projects using visual, sound and other languages that are not explicitly bound by the laws of empiricism and logic.

Acknowledgements

We thank Manuela Lacerda, Paulo Figueiredo and Pedro Cardoso, who participated in the development of the ARnatomy concept prototype, and also wrote and provided helpful comments on previous articles.

References

Adelino Cardoso, António Braz de Oliveira, Manuel Silvério Marques. [2010] *Arte Médica e Imagem do Corpo: de Hipócrates ao final do século XVIII,* (Biblioteca Nacional de Portugal [Org.], Lisboa, Portugal).

Anker, S. and Nelkin, D. [2004] *The Molecular Gaze: Art in the Genetic Age* (Cold Spring Harbor Laboratory Press, USA).

Bernardino, P. [2006] *Intersecção das novas tecnologias na criação da imagem nas artes plásticas no final do séc. XX: a imagem, a tecnologia e a arte* (Universidade de Aveiro, Aveiro, Portugal).

Bernardino, P. [2010] *Arte e tecnologia: Intersecções* (ARS vol.8 no.16, São Paulo, Brasil), http://www.scielo.br/scielo.php?script=sci_arttext&pid=S1678-53202010000200004&lng=pt&nrm=iso&tlng=pt. (accessed 7.05.2015).

Carmigniani, J. & Furht, B.. [2011] *Augmented Reality: An Overview..*" In *Handbook of Augmented Reality,* ed. Furht, B. (Springer, Berlin) pp.3-46.

Eliseu, S. [2009] *Imersão e obsessão: o carácter obsidente da produção/criação artística* (Universidade de Aveiro, Aveiro, Portugal).

Eliseu, S., Cardoso, P. [2012] *Haptic ARnatomy: Um estudo sobre possíveis convergências entre a anatomia patológica e tecnologias de realidade aumentada no campo da arte e design* (TAP - XIII Congresso Técnico de Anatomia Patológica, Figueira da Foz, Portugal), 23-24.

Eliseu, S., Monteiro, M. [2012] *Projecto ARnatomia Patológica.* (ARTECH-2012 - Conference on Digital Arts and New Media, Faro, Portugal), 385-388.

Foucault, M. [2003] *The Birth of the Clinic: Archaeology of Medical Perception*, (Taylor and Francis e-Library).

Grau, O. [2003] *Virtual Art: From Illusion to Immersion* (MIT Press, Cambridge, MA).

Kirner, C. & Tori, R. [2004] *Introdução à Realidade Virtual, Realidade Misturada e Hiper-realidade,* in *Realidade Virtual: Conceitos, Tecnologia e Tendências*, eds. Kirner, C & Tori, R., (Editora SENAC, São Paulo, Brasil), pp. 3-20.

Kirner, C. [2008] "Virtual Reality and Augmented Reality Applied to Simulation Visualization", *in Simulation and Modeling: Current Technologies and Applications,* , eds. El Sheikh, A. A. R., Al Ajeeli, A. & Abu-Taieh, E. M. O. (Hershey-NY: IGI Publishing, USA), v.1 pp. 391-419.

Latour, B. [1990] "Visualization and Cognition: Drawing Things Together", in *Representation in Scientific Practice,* eds. Lynch, M. & Woolgar, S. (MIT Press, Cambridge, MA) pp. 19–68.

Lopes, M. M., Bernardino, P. [2011] *Performando o Self no Arquivo. O Corpo Implicado: Leituras Sobre Corpo e Performance na Contemporaneidade.* (Expressão Gráfica e Editora, Fortaleza: Brasil), 95-124.

Lopes, M. M., Bastos, P. [2011] Re-enacting the Self in the Archive: Memory, Art, and Neuroscience (Isea *–International Symposium on Electronic Art*, Istanbul, Turkey).

Lopes, M. M. [2013] *Representational Strategies on Alzheimer's Disease: A Practice-Based Arts Research in a Neuroscience Laboratory* (University for the Creative Arts, Farnham, UK).

Milgram, P., Kishino, F., [1994] *A Taxonomy of Mixed Reality Visual Displays* (IEICE Transactions on Information Systems, E77-D, 12).1321-1329.

Mullen, T. [2011] *Prototyping Augmented Reality* (John Wiley & Sons, Indiana, USA).

Papagiannis, H. [2011] *The role of the artist in evolving AR as a new medium..* (Paper presented at the meeting of the ISMAR-AMH).

Huang, F., Wang, R. (eds.) [2010] ArtsIT 2009, International Conference on Arts and Technology (1st : 2009 : I-lan shih, Taiwan). Arts and technology. (Berlin: Springer, Germany).

Reichle, I. [2006] *Art in the Age of Technoscience: Genetic Engineering, Robotics, and Artificial Life in Contemporary Art* (Springer, Wiesbaden).

Reichle, I. [2003] *Where Art and Science Meet, Genetic Engineering in Contemporary Art,*
http://www.kunstgeschichte.de/reichle/pdf/artscience.pdf (accessed 7.05.2015).

Sturken, M. and Cartwright, L. [2002] *Practices of Looking: an Introduction to Visual Culture* (Second Edition). (Oxford University Press, Oxford).

7

Insights into Solid State Physics: Basic Phenomena and Teaching

David S. Schmool

Solid State Physics is a broad area of study, which, as the name suggests, refers to the fundamental physical properties of material solids. The large number of physical phenomena found in solids depends in general on the types of atoms present and their spatial distribution. The range of physical phenomena includes, crystallography, defects in solids, superconductivity, as well as the electronic, optical and magnetic properties of solids and also many subjects related to surface physics and nanotechnologies. In this paper I will give some examples of the relation between the internal structure of solids and physical properties of materials. I will also briefly address some approaches regarding the teaching of solid state physics to undergraduate students and demonstrate some basic models used to assist the understanding of physical phenomena in solid state materials.

7.1 Introduction

Solid State Physics is a broad area of physics dealing with a range of phenomena related to the material properties of solids. Of principal importance in the consideration of the physical properties of solid matter are the constituent atoms and their spatial arrangement within the solid. The spatial arrangement or crystalline structure of a solid is intimately related to the bonding mechanisms between the constituent atoms, which in turn depends on the electronic configurations of these atoms. Indeed, it is the directionality (or lack thereof) of the bonds between atoms, which will ultimately determine the equilibrium spatial arrangement of the atoms in thus forming the crystalline structure. After such considerations, we can attribute a vast majority of the physical properties of a solid to the electrons that are contained within it. Indeed, solid-state theory is a far-reaching area of physics, accounting for all electronic properties of solids, whether they are insulators, semiconductors, conductors or even superconductors. Furthermore, it is the principally the electrons in the solid which also determine their optical and magnetic properties. It is also the bonding mechanisms between the individual constituent atoms, which are responsible for the mechanical and thermal properties of the solid. It is clear then that solid state physics is directly related to all of the fundamental properties of all solids.

In addition to outlining some of the main properties of solids in this chapter, it is also intended to draw on some of my teaching experience in topics related to solid state physics and that of writing an undergraduate text on the subject [Schmool, 2014], to outline some approaches and key concepts to the understanding of the subject.

7.2 Bonding in Solids and Crystalline Structure Formation

To begin with, it is useful to consider what are the constituents of a solid, i.e. what atoms does it contain and if it is not a single element, what proportions do they exist in. Such considerations are an important starting point, since it will determine the way in which the atoms link together: this will include the strength of the bonds and their physical orientation with respect to other bonds. (We will give some examples presently.) Indeed, the linking between atoms is at the root of solid formation: no bonds, no solids or molecules! There are several mechanisms by which atoms attract one another and remain attached, depending on the ambient temperature. These all depend on the nature of the atoms, and importantly on the way in which the electrons of these atoms are arranged in their shells. A full description of the electronic states of atoms goes beyond the scope of this short chapter, (see for example chapter 1, [Schmool, 2014]), however, suffice it to say that the electrons are subject to the quantum mechanical rules of so-called *fermions*, and have well defined energy levels, which are grouped into shells. Each shell having a well-defined number of states (expressed in terms of quantum numbers). The number of electrons that an atom contains depends on the atoms and will be equal in number to the protons within its nucleus. This is what defines the atomic species. Since the proton has an equal and opposite charge to the electron, the atom is a neutral entity. (In addition to the protons, the nucleus also contains a certain number of neutrons, which are electrically neutral. The number of neutrons can vary to a small degree. Actually these form the different isotopes of an atom. The neutron serves as a "gluing" agent in the nucleus, making it more stable.) An electronic shell, having a specific number of states available, which contains a full complement of electrons is a very stable configuration. Atoms that satisfy this condition are the *inert* or *noble* gases. Indeed, so stable are they, that under normal conditions of pressure and temperature, they do not form any bonds, existing only as monatomic gases. This means that they cannot be coaxed into forming molecules or compounds.

Ionic bonding is one of the more simple mechanisms of atomic bonding. Typically it is found to occur in materials with two types of atom. A common example is the NaCl compound (rocksalt). The sodium (Na) atom has its outer electron alone in the 3s state, while the chlorine (Cl) atom has one electron missing from the 3p shell. In close proximity, an electron can be exchanged from the Na to the Cl atom, the sodium atom losing an electron to become a positive ion (Na^+) and the Cl to gain an electron to become a negative ion (Cl^-). In this way the two ions become more stable entities since their occupied electronic shells are all full. The oppositely charged ions then attract via the Coulomb force. The assembly of equal numbers of Na and Cl atoms produce the ionic crystal with a cubic structure (two interspersed face-centred cubic lattices to be exact). Figure 7.1 shows such a crystalline structure.

More complex bonding mechanisms can occur whereby an electron can be shared between atoms, as occurs with certain elements towards the centre of the periodic table. Such bonding is termed *covalent*. An example of such bonding is to be found in the diamond structure formed by carbon atoms. Each carbon atom has a valence of 4, thus allowing it to bond via this electron sharing mechanism with four other C atoms. Diamond is one of the hardest materials known to man and attests to the strong bonds that can be formed in this way. The nature of this bonding means that the spatial arrangement is such that there is an equal angle between all neighbouring bonds. This produces a tetrahedral unit, which can be understood in symmetry terms since this is the most efficient way of allowing the equivalence between all atoms involved to be taken into account in three-dimensional space.

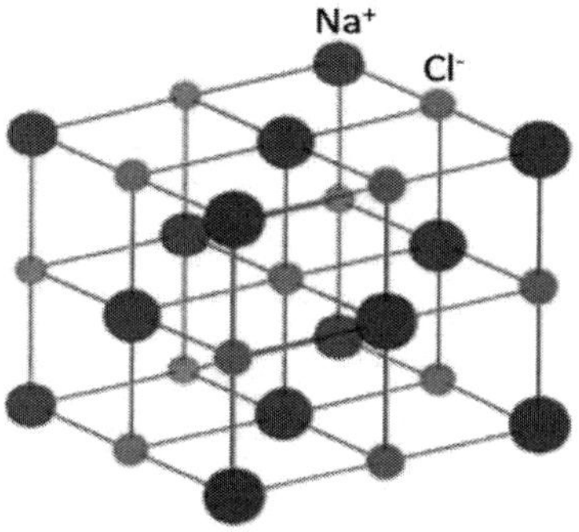

Fig. 7.1. NaCl (Rocksalt) crystalline structure.

Other compounds can be formed as a mixture of ionic and covalent bonds, such a situation is found in a number of binary systems, such as GaAs (zinc-blende structure and related to the basic diamond lattice with alternate sites occupied by different atoms) and other compound semiconductors.

Metallic systems are again different, and the free electrons that are given up by the atoms form a cohesive background for the positive ions. This allows the so-called close-packed structures to be formed, which are rather common in metals. While other forms of bonding are known to occur, they are generally weaker than those indicated above and are not typically involved in the formation of typical solids.

While this introduction may seem a little simplistic, it is a powerful method for introducing both the bonding mechanisms and explaining how crystals form with specific symmetries, due to the directionality of bonds. I find that 3D models and diagrams are an enormous aid to explanation and the understanding of these basic concepts.

7.3 Basic Concepts in the Physics of Crystals

Crystal symmetries are essential components to the understanding of the physical properties of solids, governing the mechanical, electrical, optical and magnetic behaviour of materials. The concepts of symmetry operations are a useful tool in the description of crystalline properties. The crystal is a three-dimensional array of points that are regularly separated in space. An example of a symmetry operation is generally introduced via a translational operator, which in 3D can be expressed in the form:

$$\mathbf{T}=n_1\mathbf{a}_1 + n_2\mathbf{a}_2 + n_3\mathbf{a}_3 \qquad (1.1)$$

Here the $\mathbf{a}_i$ parameters represent the minimum spacing between lattice sites in the three directions (lattice parameters) and the n_i are integers. Therefore a translation operation such as described by **T** will take us from one arbitrary position (the starting point) to another identical position in the crystal. This is what we mean by symmetry: the symmetry operation leaves us in an identical state. A full set of symmetry operations (which can include rotational, reflection, inversion and glide operations) thus describes the crystalline symmetry of a specific arrangement of points (atoms).

The fundamental base unit, which is repeated throughout the crystal by symmetry operations, is termed the *primitive unit cell.* This refers to the smallest possible unit, which must be repeated (via symmetry operations) through space to form the full crystal. In monatomic systems, the primitive unit cell will only contain a single atom. For crystals with more than one atom, the primitive unit cell will contain the *basis* and thus depends on the composition. For example, a binary compound will have a basis of two atoms; the symmetry operations relevant will then describe how this basis is repeated throughout space to form the crystal. Other unit cells can be defined and are often preferred due to convenience and the greater simplicity the can offer in defining symmetry and physical properties of the crystals concerned.

A consideration of the symmetry operations leads to the principle of the *Bravais lattice*. The lattice points of a Bravais lattice are generated by the discrete set of translation vectors defined in equation (1.1). The Bravais lattice is closed under the addition and subtraction of lattice vectors. It turns out that in two-dimensions there can exist only five types of Bravais lattice, while in three-dimensions this increases to 14.

It is frequently important in the description of crystalline solids to refer to specific directions and crystalline planes. The main convention for this is referred to via the so-called Miller indices. These are a set of three integers, *hkl*, for a three-dimensional crystal, and are used for labelling the different atomic planes and directions; the planes and directions being distinguished by the type of brackets used (*hkl*) for planes and [*hkl*] for directions, families of planes and directions are expressed using the following bracket notations respectively, <*hkl*> and {*hkl*}.

The determination of crystalline structures forms an essential part of an experimentalist's work in the characterisation of a material. The most common experimental tool to probe the crystalline structure of solids (and surfaces) is based on diffraction, typically using x-rays or electrons as the incident radiation. The wave like properties of these radiations are manifest from the conditions of constructive and destructive interference which give rise to specific diffraction patterns and are directly related to the geometry and symmetry of the crystalline lattice. The theory of the diffraction of waves in a three-dimensional crystal was developed in the early part of the 20th century by Max von Laue (1912) as well as the Braggs (father and son) a year later. While it can be demonstrated that the different approaches lead to much the same thing, it was the *Bragg law* which has received the most attention, most probably due to its relative simplicity. The law is usually expressed in mathematical form as:

$$nl = 2d_{hkl}\,\sin q_{hkl} \tag{1.2}$$

Here l refers to the wavelength of the incident radiation, d_{hkl} the interatomic planar spacing and q_{hkl} is the angle of incidence (exit). The factor n is an integer number arising from the condition for constructive interference, for the path difference for rays reflected from successive atomic layers having to be a whole number of the wavelength. The *hkl* subscripts correspond to the atomic planes involved in the diffraction process, given by the Miller indices (*hkl*). In fact for their pioneering work Max von Laue earned the Nobel Prize in Physics for the year 1914, "for his discovery of the diffraction of X-rays by crystals", while the following year the Braggs won the prize "for their services in the analysis of crystal structure by means of X-rays". Other forms of radiation are frequently used for probing the inner structure of materials, and of particular importance are neutrons. Since neutrons have no charge they are virtually non-interacting with the atomic electrons and it is from the atomic nuclei that diffraction and scattering events occur. This fundamental difference from that of x-rays and electron radiation means that additional information can be obtained, and special interest is found in the study of magnetic structures. This is possible since the neutron possesses spin angular momentum and can thus probe the spin state of the nucleus. Indeed, such information is hard to obtain for antiferromagnetic crystals, which normally do not exhibit a net magnetisation, unlike ferromagnetic and ferromagnetic materials.

As stated previously, the study of crystalline order in material solids is crucial to the understanding of their physical properties. While in most cases we treat materials as if their crystalline order were perfect, in reality we know this to be far from true. Imperfections in crystals are extremely important and can have dramatic effects on the way they perform and react under specific stimuli. In general defects in solids are characterised by their dimensionality: point defects, line defects (dislocations), planar defects and three-dimensional inclusions. All have their specific role and effect on the resulting physical properties of a solid. This concerns all physical properties (mechanical, electronic, optical and magnetic).

7.4 Electronic and Optical Properties of Solids

Electronic properties of materials are quite probably one of the most exploited in our modern man-made society, certainly in terms of the number of applications, (though possibly mechanical applications can be argued to be as important). Evidently this includes all of the electrics and electronics we use in our homes, cars and work places etc. We can also consider the enormous range of electronic devices that have mushroomed from the semiconductor industry to make modern computing, mobile telephony, gaming devices among many others, possible and that we use in our everyday lives.

In a first approach, it is useful to outline the manner in which materials can be classified, and this clearly depends on how we wish to consider the solid. For example, if we are interested in the electronic properties we generally divide materials as to whether they are good or bad conductors (insulators). The former are typically metals, with their characteristic shiny mirrored appearance, while the latter tend to be almost everything else, with the obvious exclusion of semiconductors, more of which shortly. Insulators, while not immediately apparent, are just as important in the electronics industry, for without them we would be plagued by short-circuits! Cladding on wires being an obvious example. Between these extremities of conduction are the semiconducting materials. These occupy a region of the periodic table centred around the carbon like elements of group IV in the periodic table. Extending more or less two columns either side. What is special about these materials? They have unimpressive conductive properties; though vary widely within this region of elements (also depending on the composition for alloys).

Indeed semiconductors are commonly known to be the important components on computing devices and make up the active components of integrated circuits and chips. Left in their bulk elemental state they are rather passive. However, when doped and layered they are extraordinary! In fact, the ability to vary the electrical conductivity of semiconductors is one of their prime attributes. The electrical resistivity of intrinsic semiconductors (i.e. those with no added dopants) can vary significantly with alterations in sample temperature and incident radiation. The latter has applications in photoconductivity applications (e.g. as a light sensor). Actually both cases are related; the variations of temperature and ambient radiation can be viewed as external sources of energy (i.e. thermal, k_BT, and electromagnetic energy in the form of photons, $h\nu$). The understanding of these properties of semiconductors is actually related to the way their constituent atoms bond. We mentioned in section 1.2 of this chapter that semiconductors form a mixture of covalent and ionic bonds, depending on the compounds or elements considered. While the bonds are rather strong, the electrons in these bonding or valence states are readily liberated by the absorption of small amounts of (thermal or photon) energy. The release of this electron has a dual effect. It firstly allows an electron to move into a conduction (free) states within the solid and can therefore contribute to the electrical conductivity of the material. Secondly, the empty electronic state left after the release of the electron is an unoccupied quantum state, called a *hole*, and will act as a positive charge carrier and also contributes to electrical conduction.

Further scope for manipulation of the electronic properties of semiconductors is the addition of dopants. These come in two forms: Atoms with more electrons than the host (donors) and those with less electrons available than the host (acceptors). In the former, when the dopant atoms takes up a substitutional position in the crystal lattice, the bonding occurs as with other host atoms, but now the additional electron is liberated to increase the conductivity of the sample (actually a very small amount of energy is required, but thermal energies at room temperature are usually more than sufficient to ionise all donors). Indeed, since each dopant adds an electron, the precise conductivity can be tailored via the concentration of donors; essentially the concentration of free charge carriers will be equal to the concentration of the dopant and hence the control of the samples' electrical conductivity. Semiconductors doped with donor atoms have an excess of free electrons and are referred to as *n-type semiconductors* (the *n* standing for negative, from the electron charge). A similar situation occurs with the acceptors. However, in this case the bonding removes an electron from a nearby host atom, thus creating a hole. As stated before, this also contributes to the conductivity to a level related to the concentration of the acceptors, as in the case of the donors. In this case the excess charge carriers have a net positive (*p*) charge and the resulting material is referred to as a *p-type* semiconductor. A significant majority of semiconducting devices, such as diodes, LEDs (or light emitting diodes), semiconductor lasers and transistors are formed by the formation of a junction between n and p – type materials. The p – n junction is a very important device in itself, giving the rectifying properties to the diode. This acts like a one-way valve for electrons, thus controlling their behaviour due to the inherent potential created by the formation of the junction itself.

The sophisticated modern theory of solid state physics accounts for the electronic properties of solids in terms of the quantum states that define what energies are available to the charge carriers. It is also important to understand the prevailing statistics necessary to assess the numbers of charge carriers available in each type of material. A full description of such considerations is well beyond the scope of this short chapter. The dynamics of charge carriers in crystals must also take into account the crystal symmetry itself. Such advances in solid state physics were made in the early years after the theory of quantum mechanics was developed. Such elements are the ground rules, which gives rise to the band theory of solids. This allows an energy diagram of available states to be calculated and predicts a so-called energy or *band gap* in the allowed energy states. The band gap arises from the consideration of the crystal structure and the forbidden energy states correspond to the states for strong scattering or diffraction and can be shown to correspond to the Bragg condition discussed above! This at first seems an amazing coincidence, but of course it is nothing of the kind. It is as it should be. Electrons behave as waves, as illustrated by the famous Davisson and Germer experiment of the late 1920's. So why should the intrinsic electrons of a material not be diffracted when their energies (wavelengths) correspond to a diffraction condition for the relevant interplanar crystal separation? It is actually a logical conclusion of the theory.

The band theory of solids is also applied to semiconductors, and the energy gap has some important consequences for their properties. The energy gap is also importantly the binding energy of the valence electrons that we discussed earlier. If we supply it with sufficient energy it will be released from it bonding (valence) state. Such electrons are called valence electrons and are collectively form the *valence band*. The electrons once liberated from the valence band are then free to move in the crystal (within the constraints of the crystalline symmetry) and form the *conduction band*.

It is this transition that occurs whether the energy is furnished via the thermal vibrations of the atoms or due to the absorption of a photon. In fact the band gap of a semiconductor marks a very important threshold, since this is the minimum energy that is required for photon absorption. This can be expressed as the condition for the photon energy absorption as: *hn* $\geq E_g$, i.e. the photon energy must be greater of equal to the energy of the forbidden energy gap. This allows us to then assess the given condition for the wavelength of the corresponding radiation as:

$$l \leq hc/E_g \qquad (1.3)$$

This has important implications for the optical properties of the semiconductor. Indeed the de-excitation process (i.e. that of the electron "decaying" or transiting from the conduction to the valence band, which effectively corresponds to the re-capture of a free electron by the hole state in the bonding between atoms), the electron releases an energy corresponding to the gap energy in the form of a photon with a wavelength (which corresponds to a specific colour) expressed as given in equation (1.3), where an equal sign replaces the less-than or equal symbol. Many opto-electronic devices, such as diode lasers and LEDs, function on this principle. That means the colour of the emitted radiation is predetermined by the material properties of the sample, which is related to its crystalline structure!

Developments in the 1980's in semiconducting materials related to the preparation of thin film and layered structures led to important advances in the applications of semiconductors, and can be considered as a major step forward in our capabilities to custom devices to specific working parameters. By depositing different semiconductors and controlling their composition, it is possible to tailor the band gap energy; this is known as *band-gap engineering*. This will then allow these active layers to be used to explicitly set the functioning wavelength of the subsequent device. Modern technology can do this with great precision and hence devices can be produced with good reliability.

Returning to the band-gap itself, this concept allows us to distinguish between a semiconductor and an insulator. The only fundamental difference is that the latter has a much larger forbidden energy gap, and thus at room temperature, the thermal energy is insufficient to overcome the gap. As such, very few charge carriers are available and the resulting material is insulating. Typical energies involved for semiconductors are in the 1 eV range, while for insulators are > 3 eV, we note that 1 eV = 1.602×10^{-19} J. Actually, so-called wide band gap materials have a function in extending the wavelength range available for opto-electronic devices.

On the subject of optical properties, it is worthwhile noting a few details with regards to those of metals. We are all familiar with the distinctly shiny and reflective appearance of metals. This property of metals is related to the fact that there are electrons which are essentially free to roam around the crystal (but not to escape it, there is an attractive potential which keeps the electrons inside the solid, this is called the *work function* of the material, essentially giving the energy required to remove an electron from the solid metal. This was an important factor in describing the photoelectric effect). It should be relatively clear that since the appearance of metals is mirror like, the surface of the metal must be highly reflective. In the visible region of the electromagnetic spectrum, the photon frequencies involved are low enough that the electrons in the metal can oscillate with this same frequency (i.e. absorb the electromagnetic energy) and then re-emit the energy in the form of a photon of the same wavelength. It is the electric field component of the electromagnetic wave (light) that forces the motion of the electrons. However, there is an intrinsic limit to the range of frequencies at which this can be done. The upper limit is set by the so-called plasma frequency, w_p, which can be expressed as:

$$w_p = [n_e e^2 / e_0 m]^{1/2} \qquad (1.4)$$

Here n_e refers to the electron density (number of electrons per unit volume), all other parameters are essentially constants (e is the electronic charge which has a value of 1.602×10^{-19} C; e_0 is called the permittivity of free space and has a numerical value of 8.854×10^{-12} Fm^{-1} and m corresponds to the electron mass, 9.11×10^{-31} kg). Once this frequency is reached, the electrons can no longer keep up with the variation (frequency) of the electric field and the electromagnetic wave passes through the metal, which effectively becomes transparent. From the above equation it will be noted that the plasma frequency depends on the specific value for metal's electron density. This turns out to be in the region of 10^{15} Hz, which is in the ultraviolet region of the electromagnetic spectrum. Certain metals, such as gold and copper, are notable and have distinctive colours. This is essentially due to the absorption of radiation in certain regions of the electromagnetic spectrum due to allowed transitions occurring between the localized electron energy levels on the fixed ions. For gold this occurs at an energy that is of around 2.3 eV, while for copper it is slightly lower, hence the slightly different hues of these metals. Such transition also occur in other metals, such as silver, but at higher energies which are outside the visible region of the spectrum and hence do not interfere with the reflected spectrum.

In addition to the normal electronic properties of solids there is an exceptional case of the *superconductors*. This is a fancy name and indeed a rather appropriate one. Superconductivity pertains essentially to a class of materials, which under certain conditions (principally a sufficiently low temperature) can conduct electricity without any electrical resistance. This rather exotic state of matter was discovered in 1911 by the Dutch physicist H. Kammerlingh-Onnes while studying the low temperature electrical properties of mercury. Onnes noted that mercury underwent a sudden loss of electrical resistance at around 4 K. This transition is termed the critical temperature, denoted by T_C. In elemental metals that exhibit superconductivity, the transition temperatures are extremely low, with niobium having the largest at 9.2 K. Compound materials are another matter, and the advance of oxide materials in the mid to late 1980's saw the cuprates emerge as the most promising, with YBa_2CuO_7 having a T_C of 95 K. This is significant since a cheap cryogen is available; liquid nitrogen has a boiling temperature of around 77 K. To date, the highest transition temperature at which superconductivity has been observed at ambient pressure is the compound $HgBa_2Ca_2Cu_3O_8$, with a T_C of 133 K. This has been observed to increase to 164 K under a pressure of 30 GPa. While these developments have been welcome the elusive room temperature superconductor is yet to be found.

The physical explanation of superconductivity also proved rather elusive, with the principal developments emerging only in the 1950's. The Bardeen-Cooper-Schrieffer or *BCS theory* is probably the most satisfactory in its accord with experiment and predicts a number of phenomena that are observed in this class of material. The principal component of the theory is the pairing of electrons via an interaction with the crystal lattice (electron – phonon coupling). The electrons pairs, called *Cooper pairs*, form a new quantum state, which permits them all to collapse to the minimum energy state possible, occupying the same quantum state. This is not allowed for unpaired electrons, which must obey the Pauli Exclusion Principle, meaning that the electrons cannot all occupy the same quantum and thus energy state. The normal resistive state, as found say in metals, arises from the scattering of (unpaired) electrons to different quantum states and involves transfers of energy. For the condensed quantum state of the paired electrons however, this does not occur since they remain in the same quantum state and hence move without electrical resistance. The pair state can be broken, via thermal energy, and this allows the critical temperature to enter the story, and via an applied magnetic field. Such predictions are contained within the BCS theory.

Despite this success, the BCS theory is not applicable to the so-called high-temperature superconductors. The problem is related to the high critical temperatures and the existence of the pair states. While not yet fully understood, the cuprates have CuO_2 planes and are thought to form the conducting channels in these compounds: It is evident that some form of pairing must occur, though possibly between the hole states and not the electrons. Much research effort is still directed towards this problem in solid state physics research today.

7.5 Magnetic Materials

The classification of magnetic materials is generally based on the response of a material to an applied magnetic field. The response to this field depends greatly on the internal atomic structure and crystalline structure of the sample. In the case where there are only atoms with closed electron shells, this response will be weak and will produce a small magnetisation that opposes the direction of the applied field. This is called *diamagnetism* and will occur in a majority of solids by virtue of the fact that most atoms contain at least some electrons in closed shells. When there are unpaired electrons in a solid whose magnetic moments are weakly interacting or uncoupled, the reaction to a magnetic field will be to align their moments in the direction of the applied field. Such materials are said to be *paramagnetic*.

What are generally termed *magnetic materials,* are those substances in which there is a relatively strong reaction to an applied magnetic field. In such cases there exists some form of interaction between the atomic magnetic moments that tend to align them in a specific direction with respect to one another and the principal crystalline axes. Such ordered magnetic structures can vary with regards to how the individual moments align in space. The simplest case corresponds to that in which all moments are aligned in the same direction. This is called *ferromagnetism* and is found in the transition metal elements Fe, Co and Ni as well as alloys thereof. A more complex magnetic structure occurs when adjacent magnetic moments align in anti-parallel directions. This referred to as *antiferromagnetism* and is observed for example in Cr as well as a number of iron based oxides and other compounds. The magnetisation of such materials will be zero and rather large magnetic fields are required to bring about any reorientation in the spin structure. The antiferromagnetic state is generally modelled in terms of two antiparallel sublattice magnetisations. When the sublattice magnetisations are unequal, a *ferrimagnetic* state results, with experimental properties that are similar to ferromagnetic materials, though with some important differences. More complex magnetic structures can exist in some exotic alloys, such as helimagnetism, in which the ground state is characterised by a positional variation of the magnetic moments and can arise due to competing exchange interactions. As the name suggests, the magnetic structure is helical.

The existence of a long-range magnetic order implies some form of orientational preference for the spin system. This directionality is referred to as *magnetic anisotropy*. The shape of a ferromagnetic or ferrimagnetic sample can have a very strong influence on the direction of its magnetisations due to magnetostatic energy considerations. For example, a long cylindrical sample will, in the absence of strong competing magnetic anisotropies, align its magnetisation along its long axis since this reduces the free poles at the sample surface and minimises the stray field produced by the magnetisation. Demagnetising factors can be defined and evaluated to take into account the sample shape, providing a useful tool for evaluation the effective anisotropy. Crystal field and spin-orbit interactions inside the crystal act to align the magnetic moments in a ferromagnet and give rise to the magnetocrysalline anisotropy. The crystal structure therefore can play a decisive role in the orientation of the magnetisation in the absence of applied magnetic fields. Such situations can be modelled phenomenologically and can account for the orientational dependence of the magnetisation in an arbitrarily aligned magnetic field. This process is usually performed by minimising the total energy of the system, while taking into account the various magnetic energy contributions in the solid.

The magnetocrystalline energy means that specific orientations of the magnetic moments with respect to the crystalline axes will be preferred. However, a fully saturated magnetisation comes at the cost of magnetostatic energy. It is possible to reduce this energy by dividing the magnetisation up into small regions of spontaneous magnetisation called *magnetic domains*. The formation of domains, however, also has energy costs. Only when this energy is less than the magnetostatic energy will domains form. Otherwise the sample will exist in a single domain state. The existence of domains will form regions of transition between the different domains, which are called *domain walls*. The energy of a domain wall and it thickness depend explicitly on the magnetic anisotropies and the exchange energy between neighbouring spins.

The ground state, i.e. the state of lowest energy and thus the most likely to exist, of a ferromagnet can be described as the perfect alignment of all spins along a direction on easy magnetisation. This can strictly only occur when no fluctuations of the spin system occur. Thermal energy produces an excitation of the spin system in a similar way to the production of lattice vibrations (phonons) in solids. The excitation of the magnetic spin systems produces a variation of the spin orientations called spin waves. These can be quantised (in quantum theory) as *magnons*. The existence of such magnons act to reduce the overall magnetisation of the solid and for low temperatures spin wave theory provides a good model for the variation of the magnetisation with temperature. Spin waves can be characterised as standing or travelling waves.

The motion of the magnetisation under stimulation by a fast moving magnetic field provides a sensitive measure of magnetic properties and behaviour. Precessional processes, such as ferromagnetic resonance and spin wave resonance have been used for many decades as a tool for the measurement of magnetic properties of materials. Indeed, they can furnish information on the principal magnetic properties, such as magnetisation, g-factors, magnetic anisotropies and exchange constants. Furthermore, in coupled magnetic systems, such as magnetic multilayers, it is possible to measure the interaction strength between the layers. Modern ultrafast optics techniques can be used to study real-time dynamics, such as sample demagnetisation and reversal processes in addition to the slower precessional dynamics.

The interactions of fixed magnetic moments in a solid and the free electrons can produce variations in the electrical resistivity with an applied magnetic field. This is called *magnetoresistance*. In artificially produced layered structures a giant magnetoresistance (GMR) can occur due to the spin scattering of electrons. Such processes are at the root of *spintronics*, which is a major new area of study and has produced significant progress in recent years with the existence of a number of spintronic devices, indeed the 2007 Nobel prize in Physics, awarded to Albert Fert and Peter Grünberg, was for their pioneering work in this area in the late 1980's. Much research is dedicated to the study of low dimensional effects in magnetism, such as in magnetic nanoparticles and nanostructures.

7.6 Nanotechnologies and the Extension of the Principles of Solid State Theory

Nanoscience is a very broad area of study that comprises many areas of modern science. These include all the major natural sciences of Physics, Chemistry and Biology. The principal theme is that these subjects are brought together by the common aspect that we are dealing with the properties and effects brought about by using objects of reduced physical dimensions, which as the name suggests is of the order of nanometres. The specific region of interest depending on the physical effect that is being measured. In this section we will only consider those effects with a direct connection to the physical properties of materials with size reduction in one or more dimensions.

One of the main reasons why objects show very large departures of their normal bulk properties at the nanoscale is that they have a significant proportion of their constituent atoms in surface positions. That is they are located on the surface of the object. This means they are subject to less than full atomic coordination or some modification of their normal bulk crystalline symmetry. Such modifications of the local (crystalline) structure will be accompanied by alterations in most if not all of their normal bulk physical properties. This is a rather complex physical problem, since the atoms can behave in radically different ways than when they have the normal full atomic coordination. This is because, having less bonding connections with other atoms they will have a different number of electrons available. Since the atomic electrons have a leading role in many physical properties, it is to be expected that such a situation will modify their apparent physical behaviour. In addition to this, the excess electrons may form stronger bonds with neighbouring atoms and give rise to a change in the surface crystalline ordering. Furthermore, the confinement effects produced by limiting the spatial extent of electronic wave functions can also radically alter the electronic and optical properties of solids. We will outline some of these effects in the following.

Surface modifications of crystalline order arise from the fact that the reduced symmetry of atomic coordination means that the normal bonding, which occurs in the bulk of a crystal, is incomplete. This will leave free or *dangling* bonds, which may be accommodated by a rearrangement of the surface atomic order and a relaxation of the normal interatomic separation between atoms. Such modifications of surface order is referred to as *surface reconstruction*. There are a number of specialised characterisation techniques that are only concerned with surfaces. Electron beams are of particular importance since their energies can be well controlled and from the fact that electrons in a certain energy range (20 - 1000 eV) have a very limited mean free path in solids, of the order of 1nm or less, meaning that they only probe the surface atomic layers of the solid. Such electrons can be used for diffraction purposes to evaluate surface crystallography and reconstructions as well as the electronic environment of the surface via electron spectroscopies. Actually higher energy electrons can be used to probe surfaces if their angle of incidence is reduced to a grazing incidence of a few degrees, such as in the reflection high-energy electron diffraction (RHEED) technique. Furthermore, surface microscopy techniques such as scanning electron and tunnelling microscopies and other surface probe methods have been developed over the last two to three decades. These are used in a number of ways to probe surface atomic order and morphologies as well as surface chemical mapping.

There are a large number of techniques that can be used to produce nanostructures. These go from very sophisticated methods for controlling electron and ion beams to the very basic in the form of chemical and electrochemical methods for making agglomerates of atoms and nanosized templates as well as self-assembly techniques. These technologies are broadly based on the top-down or bottom-up approaches to the fabrication of nano-systems. At the more advanced end of the spectrum, electron beams can be controlled to form very specific patterns in a photoresist, which will then be used to manufacture precisely architectured objects. Focused ion beams, or FIB, can also be used in a similar fashion to etch and deposit materials in almost exact constructions at the nanoscale. Such advanced tools are mainly used as a research technique and industrial production usually requires more large-scale methods for producing batches of devices. Replication technologies are showing themselves to be of great interest as are important developments in chemical methods of production of nanoparticles. Other forms of low dimensional structure, such as carbon nanotubes and graphene are also of enormous interest since they involve a single or multiple of single atomic layers. These materials have a range of exceptional physical properties and are strong candidates for many electronic and nano-mechanical devices and applications.

There has been a large amount of interest in micro-electro-mechanical systems or MEMS over recent decades. These are basically small mechanical devices that can be controlled or adapted to measure physical properties on the micron scale. On the other hand, they can also be manufactured to perform specific mechanical tasks, such as micromotors. The use of MEMS devices are typically designated as sensors or actuators. The formed being a detection application, while the latter is a device designed to perform an operation. In either case, the basic principle refers to the conversion of an electrical signal to a mechanical deformation of the structure or vice versa. MEMS are very popular devices since, being of micrometric dimensions, they can be built on microchip systems and incorporated in the usual CMOS microchip technologies. Such devices are common in the market place, with applications such as accelerometers and gyroscopes as well as an enormous range of sensors and actuators. Such technologies are commonly found in smart phones and gaming devices, which require position and motion sensors. The extension of technologies to the nanoscale has led to interest in nano-electromechanical systems or NEMS. While the range of applications is still limited at the current time, advances in production and characterisation are progressing. It should be remembered that the reduction of size generally leads to faster operation or higher frequency processing. Typical resonance frequencies of mechanical objects in this range of sizes can be well in to the GHz region, which is the operational frequency, used in communications systems for example.

In the area of electronics, we have seen that there is enormous scope for development. Clearly the electronics industry is one of the most important global industries and developments in nanosystems are leading to smaller and faster products. At the time of writing, the active components in transistors are reaching the scale of a few 10's of nm in size. With developments in nanotube technologies and molecular electronics, this trend can be expected to continue for the next few decades at least. At the root of these advances in electronics are the radical changes to the electronic properties that occur on the nanoscale. For example, with nanometric dimensions, insulators can become transparent to electrons and conductors can become non-conducting if there are charging effects that lead to the so-called *Coulomb blockade*. However, understanding the basic principles of these limitations allows us to manipulate the devices to our advantage. We can use our knowledge of this behaviour to produce a new range and class of device that is based on the control of the passage of single electrons. Since such devices require such small currents, there will be much less dissipation of energy, which is also an important consideration. Due to such highly controlled operation, the signal intensity can be adjusted to a very high degree using the control of the frequency of operation. This can be seen from the basic unit of electrical current, which corresponds to the number of charge carriers (such as electrons) passing through the device per unit time ($I = nq/t$). Since frequency, n, has the units of inverse period (time), a single electron device will have an operational current, which can be expressed as: $I = e/t = en$. One of the difficulties of these devices is their reliance on an energy barrier to limit the unwanted flow of charge carriers. These barriers can be relatively small, meaning that thermal effects have to be controlled and in many cases this quantum behaviour can only be observed at very low temperatures.

The energy quantisation that we observe for objects of reduced dimensions has been a property that has been well studied since the inception of quantum well technologies from the 1980s. We can manipulate the dimensions of an object to produce specific well-defined energy levels for the electronic states of a system. Such properties depend on the confinement of electrons to within the structure of the nano-object. For example a metallic nanoparticle in a non-conducting matrix will confine the electrons to within the nanoparticle itself. This allows us to control and design devices for specific applications and will have electronic and optical properties to suit specific functions. The spatial dimensions are designed to produce specific energy quantisation. This means that the electrons confined in such a structure can only have discrete energy states available to them. Therefore, we can control the colour of light emitted from semiconductor lasers and LEDs for example. The colour of light being defined by the energy difference between these energy states, which we can express in terms of the wavelength of the light emitted: $l = hc/\mathrm{D}E$, where h is Planck's constant, c the speed of light and DE the energy difference between the adjacent energy levels. Further to these properties, the optical properties of solids can be adjusted by making periodic structures, which will limit the propagation of light at specific wavelengths. Such considerations have led to the development of *photonic crystals*, which have a large range of applications even beyond the visible spectrum and can be used for camouflage at say radar frequencies for *stealth* like applications.

As with electronic properties, the magnetic behaviour of nanostructures can alter significantly from bulk materials. The magnetic characteristic length scales also fall within the nanometre region. For magnetic structures of such dimensions, there is generally not enough room to form magnetic domains and a first approximation of the nanosystem can be envisaged as a coherent macrospin. The behaviour of the magnetisation then depends on the effective anisotropy of the structure and the applied magnetic field strength and direction. The so-called Stoner – Wohlfarth model appears to be a good starting point for modelling this type of structure. This considers the magnetic entity as a single spin with a specific uniaxial magnetic anisotropy in the presence of an applied magnetic field. However, thermal agitation (as we also saw for the single electron devices), in which the magnetisation can overcome the energy barrier between energy minima as defined by the size of the magnetic anisotropy, can lead to a fast switching of the magnetic state between these energy minima of the anisotropy and gives rise to the so called *superparamagnetic* behaviour. Here there is no well-defined orientation of the magnetisation in the system. The name superparamagnetic is used because there is a similarity of the magnetic response to that of the paramagnetic state. However, in this case instead of a single spin, the ferromagnetic entity has a large number of spins which act in unison and hence the "super". The speed at which we measure is an important factor in how the magnetic response appears. For slow measurements, such as magnetoemtry, the magnetic state can oscillate between minima a number of times, while a fast measurement, such as a resonance technique, can be sufficiently rapid that the measurement is made with the magnetisation in only one of the possible states. This is a vital component in the study of superparamagnetic effects. Cooling the object to low temperatures leads to a freezing of the magnetic state, with the transition occurring at the so-called blocking temperature. These properties depend critically on the strength of the magnetic anisotropy and the size of the nano-object.

Developments in thin film and structuring technologies have followed to a certain extent a parallel path to that of electronic properties. Multilayer systems also allow the magnetic properties to be tailored via the control of exchange interactions between ferromagnetic layers and structures. In elongated and uni-dimensional structures, magnetic domain effects can be observed and exploited to make specific device applications. Of great interest are magnetic data storage technologies, much of which is based on magnetics and magnetic materials both for storage purposes and reading applications.

While we have only considered the physical response and behaviour of nano-objects, there are a large number of applications that go well beyond the physical nature of materials and utilise these properties for other purposes. Of particular importance are the many applications that are emerging in the areas of medicine and biology. Much of this is related to the use of nanoparticles, which can be exploited to label and identify regions in biological systems. For example, magnetic nanoparticles can by functionalised to attach themselves to specific proteins such as those on cancer cells. Hyperthermia treatments can be then applied locally to heat and kill the cancer cells using radio-frequency signals. Also there is much interest in using magnetic particle as a delivery system for drug therapies. Since the magnetic particles can be directed to the region where treatment is required, dosage levels can be significantly reduced, thus limiting any undesirable side effects, such as those experienced in chemotherapies. Again using magnetic particles to attach themselves to specific biological entities can allow them to be separated from the body via magnetic filtering.

We have only been able to really see the tip of the iceberg and nanotechnologies have enormous prospects for science and technology in the foreseeable future. We can expect devices to become smaller and faster and even cheaper. However, there are physical size limits and we appear to be coming close to those in a number of areas, such as in electronic and magnetic systems. This will require new thinking and future scientists will be faced with these challenges. However, this is the role of science and technology in general.

7.7 Conclusion

In this short chapter we have only been able to give a brief overview of the very broad topic of solid state physics. As will be seen, there is an enormous range of phenomena and a very large number of practical applications that have emerged from this very active area of study. Indeed, it would be no exaggeration to say that solid state physics research (and related topics) commands a significant number of practitioners and a substantial proportion of state funded research in universities and other public institutions. Its impact on industry should not be under-estimated. In fact it is difficult to find many practical devices that are not in some way related to the subject.

In scientific terms, solid state physics is the branch of physics which treats the physical properties of materials at its most fundamental level, relying heavily on quantum theory to describe the varied phenomena that are encountered. The history of science has shown how technology has progressed hand-in-hand with the development of scientific theories and experiments. Once quantum mechanics was established in the late 1920's and early 1930's, solid state physics emerged as a major area of research with spectacular successes in the understanding of materials and the development of electronic devices, with semiconductors being a particularly important example. Indeed, success has followed success and much of the major advances in our modern technology owes its origin to the progress of solid state physics, as can be seen for example in the recent advances in nanotechnologies and graphene research.

As has been outlined in this chapter, solid state physics has an important role to play in our understanding of natural phenomena. In the teaching of the subject, it is always important to outline the fundamental importance of the relation between the atomic species present in a solid and their atomic environment. These two factors are possibly the most important in defining the number of electrons (and charge carriers) available in the solid and how they are able to move through the solid under the influence of applied electric and magnetic fields.

References

Ashcroft N. W & Mermin N. D., [1976] *Solid State Physics*, (Saunders College, Philadelphia).

Bandyopadhyay S. [2012] *Physics of Nanostructured Solid State Devices*, (Springer, New York).

Blakemore J. S. [1985], *Solid State Physics*, (Cambridge University Press, Cambridge).

Kittel C., [2005] *Introduction to Solid State Physics*, (8e J. Wiley and Sons, USA).

Lindsay S. M. [2010] *Introduction to Nanoscience*, (Oxford University Press, Oxford, United Kingdom).

Myers H. P., [1998] *Introductory Solid State Physics*, (Taylor and Francis, London).

Schmool, D. S., [2014] *Introductory Solid State Physics: From the Material Properties of Solids to Nanotechnologies*, (The Pantaneto Press, London).

Wolf E. L. [2006] *Nanophysics and Nanotechnology: An Introduction to Concepts in Nanoscience*, (2e Wiley - VCH).

8

A History of the History of Gender & Science

Anne-Sophie Godfroy

This paper will present a short history of the research field "gender & science": the topics it addresses how it contributed to redefine history of science and highlighted new methodologies. The three research topics we propose to identify in a first approach, both chronological and logical, appeared one after another, but the research work is going on and all topics are now developing simultaneously. The first task of historians was to make women visible in history of science. The works of Margaret W. Rossiter [Rossiter, 1982, 1998, 2012] have been a landmark. Beyond the biographic work and the emergence of new portraits of women scientists, the role of technicians, persons without permanent positions, spouses, the correspondence, and other forms of less official and visible activities has been re-evaluated. Making women visible has encouraged a deeper understanding of the impact of gender relations in the making of science. Rossiter's paper about the "Matthew Matilda Effect in Science" [Rossiter, 1993] has been essential to understand why women were invisible. The "standpoint theory" has underlined how the belonging to a dominant or to a marginalized group can affect viewpoints and communication in science, challenging a supposed objectivity. Furthermore, through the notion of "strong objectivity", Sandra Harding [Harding, 1991] has demonstrated how the perspectives of marginalized persons can help to contribute to a more objective science. A new research field emerged in the last years. The aim is "to employ sex and gender analysis as a *resource* to create new knowledge and technology" and to propose "new practical methods of sex and gender analysis for scientists and engineers" and "case studies as concrete illustrations of how sex and gender analysis leads to innovation" ["Gendered Innovations in Science, Health & Medicine, Engineering, and Environment", 2011]. We will illustrate each topic with examples and case studies. Our aim is to provide some orientation in the growing research field of gender & science, we do not pretend to be exhaustive at all.

8.1 Introduction

The aim of the paper is to reflect on a short history of the research field "gender & science". First, we try to explore the topics it addresses; second, we show how it contributes to redefine the history of science and to question the definition of the science itself; third, we examine the methodological and epistemological issues it raises. In this paper, we mean by "science" all kind of scientific knowledge and all academic disciplines including social sciences and humanities.

We propose a both chronological and logical first approach in order to sketch some overall trends and moments and to allow the emergence of philosophical issues. The moments we describe appeared one after the other - this is the chronological approach - but we must remember the research is going on and does not always develop in a linear and progressive way, in real life, all the topics are developing simultaneously even if we try to identify some moments - this is the logical approach.

Here are the three moments we identify to outline some overall trends: making women in science visible, making gender in science visible, engendering science.

8.2 Making women in science visible

Here are some portraits of women scientists, will you recognize them?

The first and probably best known one is Maria Skłodowska, born in Poland in 1867 and died in Paris in 1934, twice laureate of the Nobel prize, co-laureate in 1903 in Physics with her husband Pierre Curie and Henri Becquerel for their research on radiations, and in 1911 in Chemistry for her works on polonium and radium.

Fig. 8.1. Marie Curie

The second picture represents Rita Levi-Montalcini, an Italian neurologist born in Torino in 1909 and died 2012 in Roma. After studying medicine in Italy and experiencing difficulties during the fascist period because of her Jewish origin, she worked in Saint-Louis, Missouri, after WWII and was the laureate of the Nobel Prize of medicine with Stanley Cohen in 1986 for the discovery of the nerve growth factor. After 1962, she worked between Saint-Louis and Roma where she chaired the neurobiology research centre of the Consiglio Nazionale delle Ricerche in Roma from 1962 to 1969 and the cellular biology laboratory from 1969 to 1978.

Fig. 8.1. Rita Levi-Montalcini

Fig. 8.2. Lise Meitner

Lise Meitner is one the third photograph. Born in Vienna in 1878, she died in Cambridge, UK, in 1968. Less lucky than the two other ones, she never received the Nobel prize even if her works on radioactivity and nuclear physics deserved it. She studied in Vienna with Boltzmann, but had no perspective of academic career at that time despite her excellent PhD thesis, and finally moved to Berlin in 1907 to work with Max Planck. There she met Otto Hahn, but experienced many difficulties because she was a woman. In 1912, things became easier when she was "invited" by Otto Hahn at the Kaiser Wilhem Institute founded in 1911, the ancestor of the today Max Planck Institute), first without a salary, later as Max Planck's assistant. In 1913, she became officially fellow of the KWI. In 1917, she chaired the department of physics with the same status as Otto Hahn, and they worked together on radioactivity till 1938, when she had to leave Nazi

Germany because she was Jewish. As the KWI was private, she was able to stay longer than researchers in public institutes who lost their positions in 1933. She fled to Sweden where she continued to work with Otto Hahn and Fritz Strassmann on nuclear fission by correspondence. They both published their results in 1938, but without mentioning the name of Lise Meitner because of the political context in Germany, even if she had played a major role in the research on nuclear fission. Possible military applications became a priority with the development of WWII. Lise Meitner was invited to contribute to the Manhattan project but declined, not wanting to be associated with the bomb. Later, she regretted to have been not enough engaged against Nazism and she criticized colleagues who worked for Nazi Germany even if they were against the regime as Heisenberg and Hahn. She left Sweden for England in 1960 and died in Cambridge. She received many prizes, but never received the Nobel she should have received with Otto Hahn in 1944 for their common works on nuclear fission. We will come back later on this trend to miss women scientists when attributing distinctions, even when the political situation is easier than Nazi Germany at war.

Hannah Arendt (1906-1975) is on the last picture. She is one of the few women philosophers remembered in the XXth century. Like Lise Meitner, she had to leave Germany because of the Nazism, and moved to France, where she was arrested but escaped before flying to the United States where she experienced difficult times as a refugee. She was back in Germany after WWII before leaving for the US again and beginning a late academic career in 1951 after her naturalization as American citizen. She was invited professor and gave lectures in political science in many universities, she published major works such as the trilogy on the origins of totalitarianism in 1951, she went to Jerusalem to attend on the Eichman trial and published in 1963 *Eichmann in Jerusalem : A Report on the Banality of Evil*, but the book was heavily

Fig. 8.3. Hannah Arendt

criticized. She became professor in political science at the University of Chicago in 1963. Even if the academic life was more open to women for the Arendt's generation, compared to Meitner's, it is very remarkable to see how long she had to wait before obtaining a permanent position in political science. It is also remarkable to notice the fact that the best known women philosopher of the XXth century was appointed professor of political science, not philosophy…

Those four women figures are among the most recognized, but there are many others, more invisible. It is commonly assumed that there were no women in science, which is not true. Women have been part of the science, but remained invisible or have been forgotten. We will explain why in the second part. Here are some examples of less visible and / or less famous women scientists.

Fig. 8.4. Madame du Châtelet by Maurice Quentin de la Tour

Many people know Emilie du Châtelet, a French woman scientist who lived during the enlightenment, but she was not alone. Isabelle Lemonon [Lemonon, 2013] discovered many women astronomers working with Lalande in XVIIIth century France. She analyses a corpus of scientific sources produced by women during the Enlightenment. The aim is a better understanding of the roles women assumed in the scientific community of the time and the identification of the social networks supporting them. Lemonon tries to understand if they contributed to science as producers, assistants, mediators or else, and how they obtained that position.

In Spain, Elena Serrano [Serrano, 2014; Serrano, n.d.], exposed the role of the "Junta de Damas" in late XVIIIth century Spain in the development of chemistry in a book about ladies becoming chemists. She shows how those women used applied sciences to improve prisons, textile industry, meals for the poor, artificial feeding for babies, etc. They combine agriculture, botany, chemistry, hygiene and health education, nutrition, domestic economy, handicraft and rural architecture. Through examples, Serrano shows how those women changed the practices, the values and even the image of Science and Technology. She analyses how the ideas about a feminine nature influenced their action in the Spanish public sphere. She explores how gendered categories redefined scientific ideas and women's social practices. Finally she analyses how the categories of gender redefined women's scientific ideas and social practices.

Fig. 8.6. Maria Lorenza de los Rios, Marquesa de Fuerte Hijar translating the works of Rumford, chemist

Fig. 8.5. Industry by Goya

Maria Rentetzi [Rentetzi, 2010] worked on women assistants and technicians from south east Europe working in Viennese laboratories between 1910 and 1938. They worked at the radium research institute on radioactivity where one third of the staff was female at a time where very few women are trained in physics or chemistry. She identified technicians but also independent researchers who published as much as men. How can we explain this specific situation? Rentetzi shows how socialist education reform and women emancipation led to women access to higher education. The political context and the geographical organization of space in the Mediziner Viertel coupled with the strong experimental culture also could have played a role in this exceptional situation. The fact one third of the scientists were women was noticed and interacted with the scientific work as James Chadwick, the British physicist, tried to use the social and gender characteristics of the experimenters, "Slavic women", to discredit the Viennese results in the controversy between Vienna and Cambridge.

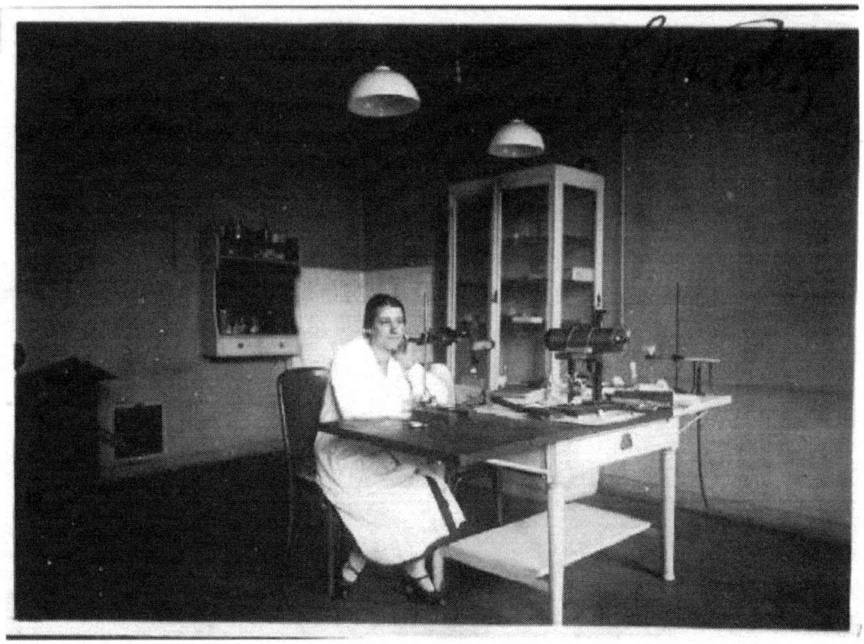

Fig. 8.7. Berta Karlik at the Institute of Radium c.a. 1925

Pnina Abir-Am [Abir-Am & Outram, 1987; Pycior, Slack, & Abir-Am, 1996] has highlighted the role played by couples and spouses in the development of science from Madame Lavoisier to contemporary science.

A major work to describe women's work in science is the pioneering work of the historian Margaret Rossiter on women scientists in America from 1880 to 1910 [Rossiter, 1982; Rossiter, 1998; Rossiter, 2012]. She found many women in astronomy, scientific employment for the federal government, in higher education, in home economics as authors of advice literature for family and home, in invisible academic positions, etc.

The history of women scientists has not only allowed a better knowledge of women scientists, it has led to new ways of studying history of science. This work has deeply renewed the sources of the history of science, and new materials and new issues emerged as new borders of science, science in the making, science in context. It contributed to transform history of science from a linear list of great discoveries but individual (male) genius into a more complex picture where networks, groups, technology, economic and political context, etc., became visible and also played a role.

Rossiter has demonstrated that, in opposition to the common idea on the absence of women in science, the more you search, the more you find.

Anyway, it is time to ask, why so few?

8.3 Making gender in science visible

A first explanation of the invisibility of women in science is the "Mathieu Matilda effect". The name was coined in a paper by Rossiter [Rossiter, 1993] in reference to the "Matthew effect" described by Robert Merton [Merton, 1968] as :

> the halo effect experienced by well-known scientists who find work attributed to themselves which they did not do.

Merton was referring to the Gospel of Matthew 13, 12:

> For to the one who has, more will be given, and he will have an abundance, but from the one who has not, even what he has will be taken away.1

The "Matilda effect" is the opposite of the "Matthew effect", it describes the fact that the names of some collaborators disappear. To sum up, the Matthew effects makes the name of the lucky collaborator always better known, he or she receives prizes, quotations, books, and his or her name is remembered, when, for the unlucky one, if the name is not totally missed by the posterity, it is seldom mentioned in the discourse of his or her colleague during the reception of the Nobel Prize or in a note in his or her bibliography. Merton attributed this unequal distribution of fame to the circular processes of accumulation of advantages: the rich get richer. We remember easier those who already benefit of charisma, previous reputation, positions in prestigious research institutions, have many former students in good positions and all those facts which contribute to bring them more and more fame and glory. The losers in the processes are often, on the opposite, the marginal people with no permanent positions, central locations or well established students able to fight for them et to protest against their exclusion, as the political know-how plays sometimes a role as important as the works themselves when creating scientific myths.

The Matthew-Matilda effect explains to a large extent the disappearance of women scientists. If there are some few success stories as Marie Curie, Maria Goeppert Mayer (Nobel Prize in Physics in 1963), or others, many others are erased. The overall description of the Matilda effect is the fact the invisible co-workers are missed. Here are a few examples: Nettie Stevens from Bryn Mawr College and Edwin B. Wilson from Columbia University discovered together the chromosomal nature of sex determination, but the discovery is attributed to Wilson alone. Frieda Robscheit-Robbins worked with the pathologist George Hoyt Whipple during thirty years and was the co-author of almost all his publications, but she did not share with him the Nobel Prize in medicine in 1934.

[1] English standard version of the Bible

The Matilda effect is not a new phenomenon; it is a constant all along the history of science. Rossiter mentions the case of Trotula, an eleventh century woman physician who lived in Salerno, Italy. She was famous for curing women's diseases. During the twelfth century, her name with misspelled by a copyist and transformed into a masculine name. When she was rediscovered by the German historian of medicine Karl Sudhoff during the twentieth century, he reduced Trotula to the role of a midwife, thinking her treatise was so important that it must have been written by a man!

The case of Rosalind Franklin is well known. She discovered with Watson and Crick the structure of the triple helix of the DNA, but she died before her collaborators won the Nobel Prize in 1962 and her role was minimized. We already mentioned the case of Lise Meitner who worked for years with Otto Hahn on nuclear fission, but was not even mentioned when Hahn received the Nobel Prize in 1944.

Fig. 8.8. Mileva Marić

Marriage was another reason to forget women scientists, and Rossiter writes that for a male scientist unsure about his academic future, to marry a bright female collaborator was the perfect strategy to neutralize a serious competitor in the race towards recognition and to gain good publications. Who remembers Mileva Marić, Albert Einstein's wife? Hertha and W.E. Ayrton worked together, but he published under his name only. Same history with Gerty and Carl Cori, when Carl alone received the Nobel Prize in medicine.

According to Rossiter, academic directories, dictionaries, histories, systematically underestimate the role of women scientists, and as a consequence, women disappear in the history of science. Hopefully, many studies are new reconstructing biographies of women scientists, and among them, Rossiter's works are very significant.

To sum up, as Johanna Russ wrote in her book *How to Suppress Women's Writing* [Russ, 1983], there are always many reasons to skip women's names in literature or in science :

> She didn't write it. (But if it's clear she did the deed. . .)
>
> She wrote it, but she shouldn't have. (It's political, sexual, masculine, feminist.)
>
> She wrote it, but look what she wrote about. (The bedroom, the kitchen, her family. Other women!)
>
> She wrote it, but she wrote only one of it. ("Jane Eyre. Poor dear, that's all she ever. . .")
>
> She wrote it, but she isn't really an artist, and it isn't really art. (It's a thriller, a romance, a children's book. It's sci fi!)
>
> She wrote it, but she had help. (Robert Browning. Branwell Bronte. Her own "masculine side.")
>
> She wrote it, but she's an anomaly. (Woolf. With Leonard's help....)
>
> She wrote it BUT. . .

In search of a name, Margaret Rossiter named this process the "Matilda effect" in memory of Matilda Gage (1826-1898) whose life was the perfect illustration of this phenomenon.

Fig. 8.9. Robert Boyle by Johann Kerseboom, 1689

A second possible explanation of the very small number of women remembered by the mainstream history of science may be the fact that modern science was constructed as a male domain, as Elizabeth Potter demonstrated in her work [Potter, 2001] on Robert Boyle (1627-1691). The famous Boyle's laws describe the relation between gas pressure and gas volume and are often considered as the perfect example of modern science based on experimentation and objective rationality. Nonetheless, when we look closer at Boyle's text, we discover many remarks hard to attribute to pure objective rationality. Science according to Boyle is universal, theoretical, rational, and those characteristics are male characteristics, on the reverse, female characteristics are local, practical and emotional ones. Boyle socially constructed antagonistic gender identities, and those gendered representations of man and women made it more difficult for women to be part of the scientific community as they were supposed to have natural characteristics opposed to the characteristics valued in modern science.

Beyond this second explanation, there is a third one: science has been sometimes totally sexist when trying to demonstrate scientifically the inferiority of women regarding intellectual abilities, as did some theories on male and female brains during the nineteenth century. As some scientific theories have been used to justify eugenics or racism, some have been used for sexist purposes, pretending having demonstrated that women were less clever, weaker, etc. All this pseudo brain science has been deconstructed by recent works on the brain physiology, as Catherine Vidal did in *Cerveau, sexe et pouvoir* [Vidal & Browaeys, 2005]. Anyway, those ideas still reappear from time to time in magazines and popular science.

Those examples show that science is never neutral and always embeds values and representations, even if scientists are not conscious of those biases. Donna Haraway [Haraway, 1990] has studied the first women primatologists. She discovered they had new ways to observe animals and were the first to observe both males and females. Male primatologists before them had observed males only and had reproduced in their observations the gendered social representations and stereotypes of their time, instead of observing objectively, as they imagined. This example is famous to demonstrate how values, racial and social representations are unconsciously embedded in scientific hypotheses, methodologies, field work, etc. There is no neutral or objective science - but this does not mean science is relativist either.

This issue leads to another question: is there a feminist science? We can mention here three possible answers. A first answer is the radical criticism of the existing science and the affirmation of a feminist science. Another answer to this question is to claim that there are no differences once we have overcome the possible stereotypes and get rid of them to go back to objectivity. There is a third answer, admitting differences but attributing them to natural differences between men and women scientists, but to their differentiated gendered socialization. This issue leads us the question of gender in science, the last part of this paper.

8.4 Engendering science

This last part is about both changing the epistemological perspectives through the introduction of gender in the science studies, and changing the contents and the methodologies of science, through the introduction of gender as an object of science and as a methodological tool for scientific analysis.

8.4.1 Engendering the subject of science

Gender studies have demonstrated how science is situated both socially and historically: science is embedded in a cultural and historical context and cannot escape the political and power challenges of the time. But if science is situated, does it mean it is relativist and there is no "good science" at all? Of course not, what is at stake is the production of scientific results less biased and less partial, this is what we call "good" science, science trying to produce the better possible knowledge about the world around us through logical reasoning and observations and experimentation when possible. Our question is how to distinguish between "bad" and "good" science if the supposed objectivity and the neutrality of science have been challenged by science studies, and especially by feminist science studies? In other words, how can we, at the same time, take into account the radical contingency of scientific knowledge and producers of knowledge and have a critical perspective on our own making of scientific knowledge? How can we be conscious of the contingency of the knowledge we produce but still try to describe and explain as accurately as we can the "real world"?

The standpoint theory is a possible answer to this difficult question. In *Whose Knowledge Whose Science* (Harding, 1991), Sandra Harding explains how we project on science our viewpoint as man or woman but also our personal viewpoint, our own situation of dominant or dominated, the ideas of our social class, our time, and how we cannot avoid this. Anyway, we are not in a relativist perspective either. She proposes the notion of strong objectivity to go beyond unconsciously biased objectivity. The idea of strong objectivity is to be conscious of our own limitation and distortions and try to overcome them through the confrontation between perspectives. Taking into account other viewpoints, as those of minorities, will provide new visions on the reality and enrich the common perspective. The viewpoint of the dominant people structures and limits what we can understand about us and about the world. The experience of dominated groups, of minorities, allows the emergence of new perspectives and new research questions, they have a strong critical potential, they are a better starting point for thinking.

When science is diverse, when it crosses different perspectives, science is more creative and innovative, but also more objective, as the results explain both the world of dominant and dominated groups. Usual conventional objectivity is not enough; the standards are not rigorous enough to detect racist, sexist, cultural biases.

Traditional "good science" is not enough. The knowing subject is supposed invisible, as knowledge is universal, distinct of what is studied, an-historical, unified and producing a unique and homogeneous knowledge. In fact, the knowing subject is incarnated and visible, immersed in a specific culture and a vision of the world, not completely distinct of the object. As the objects of knowledge are constructed, subjects are multiple and produce contradictory knowledge. Eliminating social values and political interests in science is impossible, but some values produce better results than other, less biased and partial. What is at stake is not neutrality, as it does not exist, but the values chosen by the producers of knowledge.

8.4.2 Engendering the object of science

But how can we include a gender perspective also in the object of science itself? A recent project, gendered innovations led by Londa Schiebinger et al. ["Gendered Innovations in Science, Health & Medicine, Engineering, and Environment," 2011] intend

> to employ sex and gender analysis as a resource to create new knowledge and technology

and to propose

> new practical methods of sex and gender analysis for scientists and engineers and case studies as concrete illustrations of how sex and gender analysis leads to innovations.

The project proposes several case studies in all fields. To provide some examples, one is about the design of new safety belts for cars. Sex and gender analysis of the dummies used in crash tests demonstrated that the case of pregnant women was never taken into account. The analysis led to the use of different dummies in the crash tests to study bodies of pregnant women and a greater diversity of bodies. The result was the design of a more universal safety belt, appropriate for different bodies, including bodies of pregnant women.

The study of heart disease is another case, where we can see clearly the interactions between sex and gender analysis. The case present different outcomes in the analysis. The first outcome is the redefinition of the pathophysiology of IHD:

> Analysing sex in clinical research has led to the understanding that heart disease in women often has a different pathophysiology than men, particularly in younger adults.

From this starting point, the better identification of the pathophysiology came a second important outcome:

> new diagnostic techniques are more effective than angiography for understanding the causes of IHD in women with chest pain in the absence of obstructive coronary artery disease (CAD).

This better understanding of the sex differences in symptoms allowed a third important outcome: earlier and better diagnosis of IHD in women who were before this analysis examined as if they had the same symptom as men. This better understanding of IHD allowed a better understanding of the causes of the disease in women and the specific role of oestrogens in this specific case:

> Rethinking the oestrogen hypothesis in the light of large scale trials of menopausal hormone therapy has challenged the (oversimplified) concept of cardio protective effect of oestrogens.

We see how this sex analysis may have an impact beyond the study of IHD and leads to the revision of the concepts about the role of hormones. Theoretical impact of such an analysis is not limited to the specific field where it was applied first. A last important outcome comes from the gender analysis in risk factors and prevention.

> [The case study] reveals that smoking has historically been far more common among men than women; however, in some countries, such as Sweden and Iceland, smoking rates are now higher among women (Shafey at al., 2009). The harmful effects of tobacco smoke on atherosclerosis are greater in women than in men (Tremoli et al., 2010).

This last outcome is the result of gender analysis, in other words, what is at stake is not the physical morphology but the gendered habits. Because in some countries women smoke less than me, we could assume that it is everywhere the case, but it is not, and this can modify the way we prevent and cure IHD in the different countries. Understanding both the role of physiology and the role of the gender differentiated ways of living provides a better comprehension of the combination of gender and sex effects in this type disease. From a methodological point of view, it is also a good example of the possible use of sex and gender analysis.

If these results are very promising, we must keep in mind possible misuse of sex and gender analysis. Those analyses are useful but are based on the average physiology and the average gendered habits; they are not always relevant for specific individuals. We must keep in mind that living people are not the average men or women but have their own characteristics, even if the knowledge about the average characteristics can be useful. There are sometimes more differences among the group of men or women than between men and women in general. Sex and gender analysis should not replace the attention to individual cases and make us forget that men and women have also many characteristics in common, but also other differentiated characteristics as race, social class, occupation, culture, habitation in a big city or in a rural place, etc.

The balance between being blind to sex and gender and over-emphasize sex and gender is not so easy. Being blind presents the risk of a biased knowledge which is in the fact the knowledge of the dominant group, appropriate to that group. Universality or neutrality are sometimes the other names of masculinity. But over-emphasizing sex and gender presents the opposite risk of reducing people to an average sex or gender and to assign them to predetermined roles, regardless of their effective situation.

Last remark, how sex and gender interact may be very tricky. At first glance, sex is on the side of morphological and physical characteristics, and gender on the side of habits, social roles and culture. But this easy partition cannot resist observation, nature and nurture (or nature and culture in French) are constantly interconnected. Even characteristics as height are not *natural* but depend also on nutrition, on available food, access to food, culinary culture and habits, and are, in a way, also *cultural*. At the same time, *culture* has to cope with the natural environment and resources and will develop strategies that take into account this *natural* environment.

8.7 Conclusion

To sum up our paper, we have seen how making women visible in science has changed science itself. It has changed our definition of science and brought new methodologies to study science in the making.

Making gender visible in science challenged the universality and objectivity of science itself. It makes it appear not so universal, but also full of the prejudices and values of the scientists themselves. This fact should not be considered as a negative one, but as something we cannot escape as science is part of the human history and culture, and as such, embedded in that culture. It should even be considered as positive, as it is our starting point to become conscious that science could be "better". Instead of reproducing unconscious values embedded in science, we can choose values that make science more objective and able to describe and explain the world from a large diversity of standpoints.

This leads to engendering science, introducing gender analysis in the subject of science, in other words, to encourage more diverse scientific teams to enhance creativity, but also in the object of science (as the programme gendered innovations), through new methodologies and to renew contents of science and enhance innovation.

References

Abir-Am, P., & Outram, D. (eds.) [1987] Uneasy Careers and Intimate Lives: Women in Science, 1789-1979 (Rutgers University Press, New Brunswick).

Gendered Innovations in Science, Health & Medicine, Engineering, and Environment. [2011, 2013] http://genderedinnovations.stanford.edu/index.html (accessed 9.09.2013).

Haraway, D. [1990] *Simians, Cyborgs, and Women: The Reinvention of Nature* (First Thus edition) (Routledge, New York).

Harding, S. [1991] *Whose Science? Whose Knowledge?: Thinking from Women's Lives* (Cornell University Press, Ithaca, NY).

Lemonon, I. [2013] *The invisibility of French Women'contributions to science in the 18th century : a gendered question ?* Presentation presented at the ICHSTM, Manchester.

Merton, R. K. [1968] The Matthew Effect in Science. *Science* **159**(3810), 56–63.

Potter, E. [2001] *Gender and Boyle's Law of Gases* (Indiana University Press., Bloomington)

Pycior, H., Slack, N., & Abir-Am, P. (eds.) [1996] *Creative Couples in the Sciences* (Rutgers University Press, New Brunswick, N.J).

Rentetzi, M. [2010] Genre, politique et radioactivité : le cas de Vienne la rouge. Travail, *Genre et Sociétés* **23**, 127–146.

Rossiter, M. W. [1982] *Women scientists in America: Struggles and Strategies to 1940* (Johns Hopkins University Press, Baltimore).

Rossiter, M. W. [1993] The Matthew Matilda Effect in Science. *Social Studies of Science* **23**(2), 325–341.

Rossiter, M. W. [1998] *Women scientists in America: Before Affirmative Action 1940-1972* (Johns Hopkins University Press, Baltimore).

Rossiter, M. W. [2012] *Women scientists in America: Forging a New World Since 1972* (Johns Hopkins University Press, Baltimore).

Russ, J. [1983] *How to Suppress Women's Writing* (1st edition) (University of Texas Press, Austin).

Serrano, E. [2014] *Women in the country-house: gender and rural oeconomy in late 18th century Spain.* Presentation presented at the ICHSTM, Manchester.

Serrano, E. [n.d.] *Chemistry in the city: The ladies of the Junta de damas in Madrid (1790-1812)*, n.d.

Vidal, C., & Browaeys, D. B. [2005] *Cerveau, Sexe & Pouvoir*. (Belin, Paris).

9

Philosophy, Science and Scimat

Lui Lam

Any philosophical thought system is built upon the scientific knowledge established at the time when the system is proposed. Similarly, if a philosophical system wishes to be relevant at the present time, it has to be consistent with the scientific knowledge established presently. Consequently, it is extremely important to know correctly what science really is and what the current knowledge of science is. Science is to understand Nature which includes all material systems such as humans. Scimat (Science Matters) is the new multidiscipline that treats all human-related matters as part of science, from the perspective of complex systems. Here, the nature of science, the concept of scimat, philosophy as a science matter, the proper way to integrate the humanities and "science", the fallacy of scientism, and the misconception that leads to Zeno's Paradox are discussed.

9.1 Introduction

Any philosophical system is built on the scientific knowledge known at that time. Similarly, if a philosophical system is to remain valid presently it has to be consistent with the scientific knowledge currently known. Thus, it is extremely important to understand correctly what science is and what the current scientific knowledge is.

This chapter[1] explains that "everything in Nature is part of science"; in particular, everything related to humans are part of science. And this is the basic premise of *Scimat* (Science Matters), the new multidiscipline

---

[1] The content of this chapter was presented in Chinese at the High Forum on Natural Dialectics and Advancement of Science, Renmin University of China, Beijing, May 27, 2011. In this English version, a few places and references are updated and additional footnotes are added

proposed by Lui Lam in 2007/2008 [Lam, 2008a, 2008b]. Thus, Philosophy,[2] a discipline in the humanities, is also part of science.

This chapter tries to – from the scimat perspective -- clarify the common misunderstandings about humanities and science, and discuss problems related to philosophy, the synthesis of humanities and "science",[3] and scientism. Finally, the mistakes about Zeno's Paradox are pointed out.

9.2 Science

Science is humans' serious effort to understand Nature *without* bringing in God or any supernatural [Lam, 2014], a process starting with the ancient Greeks about 2,600 years ago. And Nature includes all material systems in the universe. Presently, our understanding of the universe [Turner, 2009] is that it began with the Big Bang which occurred 13.7 billion years ago. Protons and neutron appeared 10^{-5} second later, which bind together to form nuclei 0.01-300 second after the Bang. Atoms appeared much later, 380,000 years after the Bang. Stars were formed 13.4 billion years ago (from which almost all the atoms on Earth came [Berkowitz, 2012]). The solar system which includes Earth appeared 4.7 billion years ago; life on Earth, 3.7 billion years ago. The important point is that all living and nonliving systems on Earth are formed of atoms.

The origin of life on Earth remains an unsolved problem. But we know from Darwin's evolutionary theory (1859) that humans did evolve from much simpler living systems. Additionally, we know that six million years ago, the lineages of humans and chimps split from each

---

[2] Philosophy about 2,600 years ago was the single discipline created by the ancient Greeks. It split into 'Philosophy' (about humans), Natural Philosophy (about nonhuman systems) and Theology around the 14th century. In the 18th century 'Philosophy' was further split into Humanities and Social Science. "Philosophy" (with double-quotation marks), the Philosophy commonly understood today and the one *referred to in this chapter*, is a discipline within the Humanities; it is about "metaphysical" problems and is a small subset of the ancient Greeks' Philosophy. See [Lam, 2014, Figs. 1.2 and 1.4].

[3] In this chapter "science" with double-quotation marks means science in the conventional sense, which is the sum of "natural science" and social science but excludes the humanities (see [Lam, 2014]).

other, and our kind of *Homo sapiens* appeared only 195,000 years ago [Lam, 2011, p. 6].

Consequently, all studies about humans (humanities in particular) are part of science because humans are a material system and science includes the study of all material systems. In other words, "everything in Nature is part of science" [Lam, 2008a].

Sometimes, control experiments are considered one of the characteristics of any science. Yet, for historical disciplines such as astronomy and archaeology this is not true. Similarly, if we insist that science must have the ability to predict then we will be forced to give up the early stages of any science. For example, when Thales (c. 624-c. 546 BC) proposed that everything is made of water he was not able to predict anything but we still call him the father of science.

9.3 Scimat

Knowledge in the world can be divided into two types: one type is human-independent; the other type, human-dependent. For example, Newton's three laws of motion are human-independent, meaning that they can be discovered by smart aliens, too, if they exist. Examples of human-dependent knowledge are literature and applications of science. Human-independent knowledge is those commonly called "natural science"; human-dependent knowledge, humanities and social science. However, this is not entirely correct. As described above, humans are a material system called *Homo sapiens* that is made up of atoms, the same atoms that are studied in "natural science".[4] Thus, all human-dependent knowledge should be part of natural science since the latter covers all material systems. Consequently, a logical conclusion is that

$$\begin{aligned} \text{science} &= \text{natural science} \\ &= \text{human-independent science} + \text{human-dependent science} \quad (1) \end{aligned}$$

[4] Knowing that all materials are made up of atoms does not imply that for every problem under study we have to start from the level of atoms. This is not even so in physical problems. In fact, for any discipline there always exist three different research levels (see Section 4).

Human-dependent science is called *Scimat* [Lam, 2008a, 2014][5] which is about probabilistic, complex systems.[6] Scimat includes all the topics in humanities and social science as well as medical science (especially those with bearing on the former two). Thus, philosophy—derived mostly from humans' thinking—is part of science.[7]

Human-independent science is what is commonly known as modern science which is mostly about simple systems. Modern science has progressed drastically since Galileo (1564-1642), resulting in the underdevelopment of scimat (except medical science) which are about complex systems (Fig. 9.1).

Yet the behaviour of complex systems is quite different from that of simple systems and is more difficult to study, even though there do exist three "universal" principles (chaos, fractal and active walk [Lam, 1998]) that are equally applicable to simple and complex systems [Lam, 2008a]. Moreover, the human system is most complex and is probabilistic.[8] In other words, deterministic predictions about human matters are impossible (except that every human being is predicted to die someday, but we cannot tell which day). Mistakenly, many scholars and laypeople alike identify science with the science of simple systems, and thus identify probabilistic, complex systems with deterministic, simple systems. This common misunderstanding is unfortunate and has grave consequences. Most human tragedies can be traced to this mistake.

Scimat aims to understand everything related to humans; every appropriate research methods and tools could be used, including evolutionary theory, cognitive science, neuroscience, statistical physics, complex systems science, etc. Scimat connects effectively the humanities and "science" (Fig. 9.2). From the scimat perspective, we have studied human history [Lam, 2008c] and established the new discipline

---

5 Scimat is *Renke* in Chinese, meaning Human Science [Burguete & Lam, 2013].

6 For complex systems see [Lam, 1998]; for philosophy of complexity see [Liu, 2008].

7 Karl Marx (1818-1883) was not always right. But he was right when he said: "natural science" will later include the science of humans just like science now includes "natural science"; human studies will become a discipline in science [Marx & Engel, 1979, p. 128]. See also [Liu, 2003].

8 It is probabilistic because a human being is an open system which continuously interact with its environment.

Histophysics [Lam, 2002] (see an example on Chinese history in Section 6). Additionally, we have proposed a new explanation of the origin and nature of arts—an unsolved problem in art studies for 2,400 years since Plato [Lam, 2011].[9]

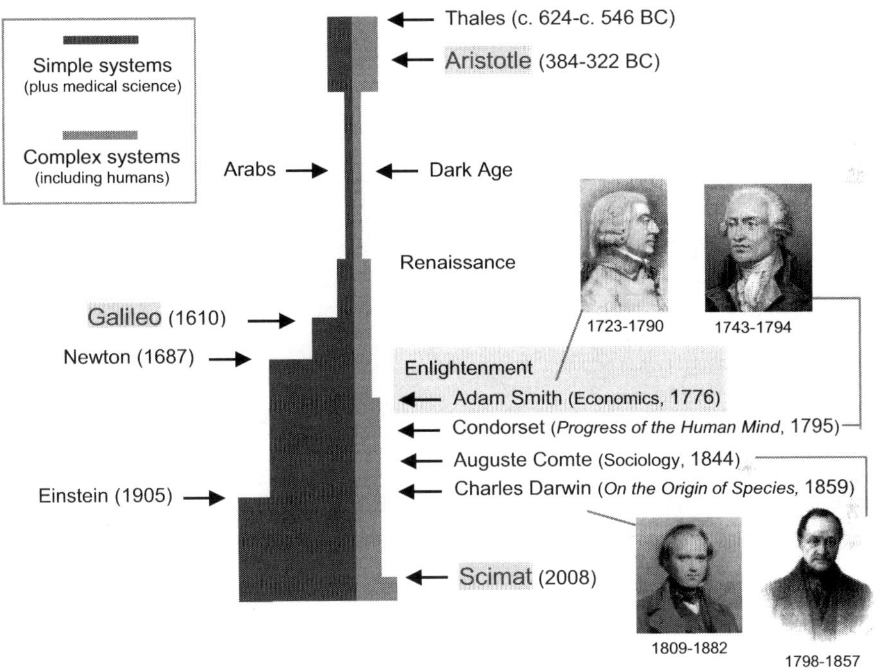

Fig. 9.1. A brief history of science in the last 2,600 years since Thales [Lam, 2011]. The left (right) column represents simple (complex) systems; the column width, roughly the achievement attained in different time period. Adam Smith established Economics, the first discipline in social science, in 1776 during the Enlightenment (1688-1789) while August Comte founded Sociology in 1844. The rapid advance of the science of simple systems resulted in the underdevelopment of the science of humanities which are complex systems, leading to humanity's many difficulties or tragedies.

[9] The Scimat Program to advance scimat was started by Maria Burguete and Lui Lam in 2007. Presently, we have (1) set up a biennial international conference series (held 2007, 2009, 2011 and 2013 in Portugal); (2) set up a scimat book series (publisher: World Scientific; founder and editor: Lui Lam); (3) set up an international scimat committee. Scimat website: www.sjsu.edu/people/lui.lam/scimat. See also [Burguete & Lam, 2016].

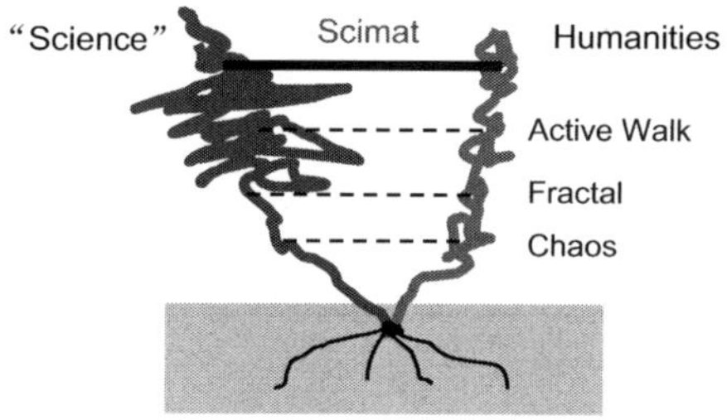

Fig. 9.2. Scimat links up completely the humanities with "science" while active walk, fractal and chaos, respectively, does that partially. Humanities and "science" share the same root, growing up like two branches of the same plant [Lam, 2011].

9.4 Philosophy: The Scimat Perspective

The word philosophy means wisdom, the wisdom about all kinds of things. Thus, at its very beginning in ancient Greece, philosophy is to understand not just humans but all things, small or big, such as whether the universe is finite or infinite in size, and the nature of time.[10] Later, when many topics originally studied in philosophy were hijacked by physics [Morris, 2002], philosophy retreated[11] and shrunk to the study of highly abstract, unsolved problems of great difficulty (truth, meaning of life, etc.) as we know it today.[12]

The correctness of scientific knowledge underlying the philosophical system is extremely important; when it is wrong, the whole system crumbles. As an example let us consider the philosophical arguments of

---

[10] The Sept/Oct 2014 issue of the magazine *Philosophy Now* focuses on the early Greek philosophers: Thales, Diogenes, Heraclitus and the "crazy Aegean gang" including Democritus. Fun to read. See [Steinbauer, 2014] for an introduction.

[11] Similarly, the Christian religion retreated when "science" advanced [Lam, 2004].

[12] For beginners, a gentle entry to "philosophy" could be Fernando Savater's *The Questions of Life* [2002]. It contains, in an Appendix provided by the translator Carolina Ospina Arrowsmith, a brief introduction to the lives and ideas of important philosophers since Socrates. A quick summary of philosophers and their ideas, arranged chronologically starting with Thales and Lao-Zi, is given in *The Philosophy Book* [Buckingham et al, 2011]. A clear and authoritative survey is *The History of Western Philosophy* by the mathematician and philosopher Bertrand Russell (1872-1970) [1945], a Nobelist in Literature (1950). An academic resource is Stanford Encyclopedia of Philosophy (plato.standford.edu).

Immanuel Kant (1724-1804). In Kant's days, *the* science is Newton's deterministic mechanics. Thus (1) he first assumed, incorrectly, that science belongs to the deterministic domain; (2) he also realized, correctly, that morality is nondeterministic and belongs to the "freedom" domain; i.e., humans have freedom of choice in morality matters. (3) Then he asked: If everyone is free to choose what he or she wants or does, how can we guarantee that world is rational and meaningful? To solve this problem, (4) he claimed: We need "religion" (or a higher principle), which will govern morality matters and bring meaning to our lives [Ye, 2007, pp. 6-9]. This series of arguments is no longer valid, judging from what we know today. Kant's philosophy, at least this part, is outdated. The reason is that we now know that science is not just about deterministic systems but is also about probabilistic systems. Moreover, even though it is not yet final, morality can be linked to the evolutionary advantages of the humans [Harris, 2010; Shermer, 2015] and could even be hard-wired in the brain [Waal, 2015]; thus, religion is not absolutely needed here. Finally, to say that life must be meaningful is an assumption slipped in by Kant.[13] Like many other philosophers, Kant's arguments are full of metaphysical presumptions and logical loopholes.[14] (See Russell [1945].)

For a long period of time, in philosophy, the mind is considered to be completely separated from the body—the so-called mind/body dualism thesis (or conjecture). In other words, thinking could not be reduced to the brain's physiochemical functions. However, this duality

---

[13] If life does not have any a priori meaning, it does not follow that life has no meaning. An individual can find meaning to her life by helping others or observing sunsets, say. The world is full of wondrous things.

[14] Here is how the historian and philosopher Robin Collingwood (1889-1943) rated Kant's works: "So long as he confines himself to drawing the distinction between philosophical method and mathematical, his touch is that of a master; every point is firm, every line conclusive. But when he turns to give a positive account of what philosophy is, his own distinction between a critical propaedeutic and a substantive metaphysics, hardened into a separation between two bodies of thought, becomes a rock on which his arguments splits" [Collingwood, 2008, p. 25]. As for Bertrand Russell, he says, "Immanuel Kant (1724-1804) is generally considered the greatest of modern philosophers. I cannot agree with this estimate, but it would be foolish not to recognize his great importance" [Russell, (1945) 1972, p. 704]. That is, Collingwood considered Kant's work on philosophy mostly wrong and Russell thought Kant was overrated.

view is no longer in fashion, for two reasons. (1) As pointed out by the philosopher Jaegwon Kim, "[I]mmaterial nonspatial minds would be totally impotent, and this renders them explanatorily irrelevant and useless" [2005, p. 151]. (2) The development of cognitive science [Kolak et al, 2006] makes it possible to connect humans' many sensations and thinking with the operation of the brain's neurological system. That is, the mind has a physical (or biological) basis. All these point to the fact that human-related matters are physically based and thus can be studied scientifically—the premise of scimat.

Like any other discipline, philosophy can be studied with three different approaches (or at three different levels): empirical, phenomenological and bottom-up [Lam, 2002, 2011]. These three approaches complement and mutually support each other. What approach to use should depend on the specific problem under study. Most of the philosophy studies are at the empirical or phenomenological level, as is the case in the early stages of any scientific study. The recent Neurophilosophy [Churchland, 1989], a branch of *Neurohumanities* [Lam, 2011, p. 22], is an example of the bottom-up approach in philosophy.

9.5 Synthesizing Humanities and "Science"

The synthesis of humanities and "science" is an important topic in both the East and West. In the West, this topic appears in the form of the "two-culture problem". (See [Lam, 2008a] for a detailed discussion.) In China, this problem was recognized early on by Cai Yuan-Pei (1868-1940),[15] and later on by Gong Yu-zhi (1929-2007) [2005] (Fig. 9.3) and others.

---

[15] Cai Yuan-Pei went abroad to study in Germany at age 40 and again in France at age 46. As President of Peking University (1916-1927), he reformed the university and started the tradition of putting "academic" and "freedom" first in a university education. In 1918 he pointed out that "university is a place purely for the pursuit of knowledge, not the place to produce professional certificates nor the place for selling knowledge. A scholar should be interested in researching knowledge and should train oneself to be a knowledgeable person with high character" (http://baike.baidu.com/view/2008.htm?fr=ala0_1_1#3, Aug. 28, 2010). In 1928 he established the National Academy of Art (now called China Academy of Art) in

In fact, Cai's understanding on this issue is very deep, more in line with that of scimat. In 1918, he already recognized that there is "science" in the humanities, vice versa; he opposed separating "science" from humanities. And he was a man of action. As the president, he reorganized Peking University into departments according to the disciplines, and reformed the curriculum [Sun, 2004]—the beginning of a truly general education in Peking University but, unfortunately, not in the rest of China.

Fig. 9.3. Cai Yuan-Pei (left); Gong Yu-Zhi (right).

In the West, with rapid advance of "natural science" in the last 200 years, humanities were underdeveloped and two cultures (humanities and "science") were formed. In 1959, C. P. Snow [Snow 2012 (1959)] pointed out the two-culture problem (in United Kingdom). To bridge the gap he suggested that each side to learn something about the other side—a strategy adopted in general education today in the West and in China. However, this method is ineffective and insufficient because it does not go to the root of the problem, which is the misunderstanding of the definition of science [Lam, 2008a]. The effective step is to start from the

Hangzhou *and* the National Academia Sinica (the foreunner of Chinese Academy of Sciences) in Nanjing. Cai was President of the Academia Sinica from 1928 to 1940; went to Hong Kong from Shanghai in 1937 due to Japanese invasion; died of illness in Hong Kong in 1940 (http://zh.wikipedia.org/zh-cn/蔡元培, Aug. 28, 2010).

fundamentals: (1) Recognize that humanities are part of science; (2) let everybody know about the three common principles governing both sides, viz., chaos, fractal and active walks [Lam, 1998, 2005]; (3) raise the *scientificity* (scientific level) of the humanities by encouraging collaboration between humanists and "scientists" (as advocated by scimat). Better, teach everybody the unified perspective of humanities and "science" through a new general-education course [Lam, 2017].

9.6 Scientism

There are two definitions of scientism: (1) Science is almighty; (2) the so-called Scientific Method can be used in all research disciplines [Gong, 2005, p. 384]. But both these two assertions violate the scientific spirit, which is to keep an open mind. No conclusion or method is valid before it is done and proven to be "true" or applicable. Scientism could be coming from people's experience in dealing with simple systems but the world consists of not just simple but complex systems, too.

As an example, let us consider the distribution of lifetime of Chinese dynasties, from Qin to Qing. As shown in Fig. 9.4, historical laws do exist (at least in this case). The lesson is that the existence of historical laws usually cannot be predicted; their existence is ascertained only when they are found. What is needed is that someone has to go out and look for them, by assuming (even just temporarily) that they do exist.

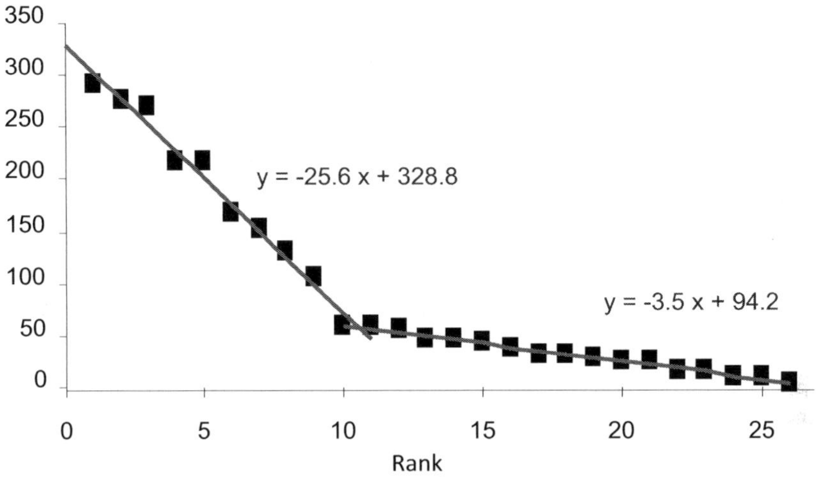

Fig. 9.4. Distribution of lifetime of Chinese dynasties, from Qin to Qing (221 BC-1949) [Lam, 2006]. The dynasty lifetimes are arranged from large to small, from left to right—the Zipf plot. Only one of the overlapping data is kept. The result shown here is a particular example of the "universal" bilinear effect [Lam et al, 2010]. The historical law exemplified here is that (1) the dynasties lifetimes are discrete; (2) a dynasty can last another 3.5 years once it survives 3.5 years; but once it passes 57 years, it can survive every 25.6 years. It also predicts that the dynasty after Qing, if exists, will end exactly in its 329 year if it lives longer than Tang. The "curse of history" indeed exists in the case of Chinese dynasties.

In scimat we say only that "everything in Nature is part of science". We insist that the researcher in any discipline has to keep an open mind, be serious, be honest, and is willing to admit mistakes [Lam, 2008a]; i.e., in the words of Gong Yu-Zhi [2005, p. 372], "be realistic". And this is the well-tested scientific approach. We have not insisted that a particular research method would lead to success. Thus, scimat is *not* scientism. We welcome anybody to use her own method of choice to do her research. The scientific spirit is tolerant and pluralistic, with the reality check as the final judge [Lam, 2014]. Only the funding agencies have to decide who they want to support.

Yet, it should be pointed out that a good research method should be able to solve a large number of problems. The scientific approach or process is the one we know of that was capable of doing this, and the *only* one that enabled us to make cell phones, say.

9.7 Zeno's Paradox

It is beneficial for philosophers to know a little bit more science (especially that of complex systems) and mathematics. For instance, this could prevent them from making the mistakes committed by Zeno of Elea (c. 490-430 BC), the early Greek philosopher. The birth of Zeno's Paradox was due to his deficiency in mathematics. He mistakenly thought that an infinite series (i.e., the sum of an infinite number of numerical terms) must be infinite—not his fault though, since infinite series was not well studied at his time.

The Zeno's Paradox, one of his many paradoxes, can be expressed in more than one way. The simplest version is related to the problem raised in *Zhuang-zi: The Society Chapter*. In this Chapter, Zhuang-zi (c. 369-286 BC) remarks: A one-foot long stick, take half each day, and you will never exhaust it. Here Zhuang-zi did not know that a stick is make up of discrete atoms and viewed it as a continuous body. Similarly, Zeno remarked that a man could not move from point A to point B because, starting from point A, he had to move to the half point first, then the half point of the rest distance, ad infinitum; and because there is an infinite number of steps, he could never reach point B. What is wrong with this argument, mathematically speaking?

Even if we accept Zeno's presumption that this person is a point mass with zero size and the distance from A to B is a continuum, his arguments are still wrong for the following reasons. Let the distance from A to B be L, the person moves with constant speed v, and the total time traveling from A to B be t. Then

$$t = t_1 + t_2 + t_3 + \ldots \tag{2}$$

Here t_1 is the time moving from A to the half point between A and B, t_2 the time to cross half of the remaining distance, etc. It is easy to see that $t_1 = \frac{1}{2} L/v$, $t_2 = \frac{1}{4} L/v$, etc. Equation (2) becomes

$$t = L/v\,(1/2 + 1/4 + 1/8 + 1/16 + \ldots) = L/v \tag{3}$$

That is, the total time of travel is L/v, a finite number, even though there is an infinite number of terms in Eq. (3).[16] The fact that the sum of an infinite number terms is finite is because each term decreases fast enough. Zeno's mistake is that he thought that the sum of an infinite number of terms must be infinite ($t = \infty$).

Zeno was already wrong at the mathematical level. But when he and other philosophers (and laypeople) applied his argument to the real world, they were wrong by ignoring the fact that everything, the walker or the stick, is made of atoms and is not a point particle. The thing has a finite size and the Zeno procedure cannot be carried out by real objects in practice [Gong, 2005, pp. 138-140; O'Carroll, 2014].

9.8 Conclusion

Ultimately, all serious and honest pursuit of knowledge is to understand Nature. But Nature includes two kinds of material systems: the human system and nonhuman systems. They are all legitimate objects/subjects in research. Thus all of these researches are within the domain of science even though the method used may be different. For example, literary scholars use their bodily sensors and their brain as information processor while "natural scientists" use, additionally, measuring instruments and computers. Yes, all these activities can be view from a unified perspective: They are scientific developments at various stage of maturity. There is a lot that they can share with and learn from each other.

Many scholars, from Aristotle (384-322 BC) to da Vinci (1452-1519), are fully aware that "everything in Nature is part of science". Yet, not until recently with the successful experience gained in modern science—from evolutionary, cognitive and neuro- sciences, statistical physics, complex systems science and other disciplines—that we know how to study human-related matters scientifically. Scimat follows the

[16] Here an algebraic proof is given. Let $x = 1/2 + 1/4 + 1/8 + 1/16 + \ldots$ Then $x = ½+ ½(1/2 + 1/4 + 1/8 + \ldots) = ½+ ½x$. Thus $x(1 - ½) = ½$, resulting in $x = 1$. QED.

Aristotelian tradition of treating all human-independent and human-dependent matters, philosophy included, as part of science.

We concur with Bertrand Russell when he writes:

> Most philosophers...profess to be able to prove, by a priori metaphysical reasoning, such things as the fundamental dogmas of religion, the essential rationality of the universe, the illusoriness of matter, the unreality of all evil, and so on....This hope, I believe, is vain." [Russell, (1912) 2010, p. 99]

But we beg to differ when he claims that

> Philosophy is to be studied, not for the sake of any definite answers to its questions, since no definite answers can, as a rule, be known to be true, but rather for the sake of the questions themselves. [Russell, (1912) 2010, p. 112]

This was definitely not the attitude of the ancient Greek philosophers, the pioneers, nor the late comers like Kant, even though it could be the attitude shared by many contemporary philosophers when they are left to deal with the most difficult problems not yet hijacked by the "natural scientists". The prospect of never reaching the truth (if true) never deters the "natural scientists" from looking for an answer, even only a provisional answer. And there is no reason to expect less from the philosophers. In this regard, raising the scientific education of philosophy students and combining theoretical approach with empirical approach in philosophical research (see, e.g., [Churchland, 1989; Churchland, 2007; Knobe & Nichols, 2008]) are two things that should be encouraged.

References

Berkowitz, J. [2012] *The Stardust Revolution: The New Story of Our Origin in the Starts* (Prometheus Books, New York).

Buckingham, W., Burnham, D., Hill, C., King, P. J., Marenbon, J. & Weeks, M. [2011] *The Philosophy Book: Big Ideas Simply Explained* (DK Publishing, New York).

Burguete, M. & Lam, L. (eds.) [2013] *Renke: Humanities as Complex Systems* (Renmin University of China Press, Beijing).

Burguete, M. & Lam, L. [2016] "The Scimat Program: Towards a better humanity", in *Humanities as Science Matters: History, Philosophy and Arts*, eds. Burguete, M. & Riesch, H. (Pantaneto Press, Luton, UK).

Churchland, P. [2007] *Neurophilosophy at Work* (Cambridge University Press, Cambridge).

Churchland, P. [1989] *Neurophilosophy: Toward a Unified Science of the Mind-Brain* (MIT Press, Cambridge, MA).

Collingwood, R. G. [2008] *An Essay on Philosophical Method* (Clarendon Press, Oxford).

Gong Yu-Zhi [2005] *Natural Dialectics in China* (Peking University Press, Beijing).

Harris, S. [2010] *The Moral Landscape: How Science Can Determine Human Values* (Free Press, New York).

Kim, J. [2005] *Physicalism, or Something Near Enough* (Princeton University Press, Princeton).

Knobe, J. & Nichols, S. (eds.) [2008] Experimental Philosophy (Oxford University Press, Oxford).

Kolak, D., Hirstein, W., Mandik, P. & Waskan, J. [2006] *Cognitive Science: An Introduction to Mind and Brain* (Routledge, New York).

Lam, L. [1998] *Nonlinear Physics for Beginners: Fractals, Chaos, Solitons, Pattern Formation, Cellular Automata and Complex Systems* (World Scientific, Singapore).

Lam, L. [2002] "Histophysics: A new discipline", *Modern Physics Letters B* **16**, 1163-1176.

Lam, L. [2004] *This Pale Blue Dot: Science, History, God* (Tamkang University Press, Tamsui).

Lam, L. [2005] "Active walks: The first twelve years (Part I)," *Int. J. Bifurcation and Chaos* **15**, 2317-2348.

Lam, L. [2006] "Active walks: The first twelve years (Part II)," *Int. J. Bifurcation and Chaos* **16**, 239-268.

Lam, L. [2008a] "Science Matters: A unified perspective", in *Science Matters: Humanities as Complex Systems*, eds. Burguete, M. & Lam, L. (World Scientific, Singapore) pp. 1-38.

Lam, L. [2008b] "Science Matters: The newest and biggest transdicipline", in *China Interdisciplinary Science*, Vol. 2, ed. Liu, Zhong-Lin (Science Press, Beijing) pp. 1-7.

Lam, L. [2008c] "Human history: A Science Matter", in *Science Matters: Humanities as Complex Systems*, eds. Burguete, M. & Lam, L. (World Scientific, Singapore) pp. 234-254.

Lam, L. [2011] "Arts: A Science Matter", in *Arts: A Science Matter*, eds. Burguete, M. & Lam, L. (World Scientific, Singapore) pp. 1-32.

Lam, L. [2014] "About science 1: Basics—knowledge, Nature, science and scimat", in *All About Science: Philosophy, History, Sociology &*

Communication, eds. Burguete, M. & Lam, L. (World Scientific, Singapore) pp. 1-49.

Lam, L. [2017] *Humanities, Science, Scimat* (World Scientific, Singapore).

Lam, L., Bellavia, D. C., Han, X.-P., Liu, C.-H. A., Shu, C.-Q., Wei, Z., Zhou, T. & Zhu, J. [2010] "Bilinear effect in complex systems", *Europhysics Letters* **91**, 68004.

Liu Da-Chun (ed.) [2003] *Development of Humanities and Social Science at Renmin University of China, 2002* (Renmin University of China Press, Beijing).

Liu Jin-Yang [2008] *Complexity: A Philosophical View* (Hunan Science and Technology Press, Changsha).

Marx, K. & Engels, F. [1979] *The Complete Works of Marx and Engels*, Vol. 42 (People's Press, Beijing).

Morris, R. [2002] *The Big Questions: Probing the Promise and Limits of Science* (Henry Holt, New York).

O'Carroll, F. [2014] "Gerry's paradox", *Philosophy Now*, Issue 104, Sept/Oct, 53-54.

Russell, B. [(1912) 2010] *The Problems of Philosophy* (Simon & Brown, www.simonandbrown.com).

Russell, B. [1945] *The History of Western Philosophy* (Simon & Schuster, New York).

Savater, F. [2002] *The Questions of Life: An Invitation to Philosophy* (Polity, Cambridge, UK).

Shermer, M. [2015] *The Moral Arc: How Science and Reason Lead Humanity toward Truth, Justice, and Freedom* (Henry Holt and Company, New York).

Snow, C. P. [2012] (1959) *The Two Cultures* (Cambridge University Press, Cambridge).

Steinbauer, A. [2014] "Step out of my sunlight!", *Philosophy Now*, Issue 104, Sept/Oct, 4.

Sun Xiao-Li [2004] *The Synthesis of Humanities and Science: A Scientific Trend in the 21st Century* (Peking University Press, Beijing).

Turner, M. S. [2009] "The universe", *Scientific American*, Sept., 36-43.

Waal, F. B. M. de [2015] "Hard-wired for good?", *Science* **347**(6220), 379.

Ye Xiu-Shan [2007] "Philosophy as the spiritual home", in *General-Education Lectures: Philosophy (I)*, eds. Lu Ting & Xu Hong (Culture and Art Publishing House, Beijing) pp. 1-22.

10

Humour, Science and Seriousness

Hauke Riesch

Science, like any other area of social interaction, abounds with humour and humorous remarks, despite the outward appearance of science as a serious business. While studies have previously looked at scientists' humour in informal situations, this chapter will try to give a brief account of humour appearing in serious scientific publications. These will be examined through three case studies that highlight some of the potential social norms, values and consequences of humour that transgresses the otherwise clear norms of scientific writing.

10.1 Introduction

Humour within and about science is not a topic that has attracted sustained academic attention. One reason for this is that science has traditionally cultivated a very serious and even humourless image, and therefore it is often perceived to be not a proper subject for humour to occur naturally. Another reason probably is that humour and comedy are themselves not very developed areas of social scientific attention: almost every classic sociological monograph on humour starts with an apology that such a light-hearted topic needs serious and weighty academic tomes devoted to them. Yet the need is there; as these works cogently argue, humour is a fundamental and widespread aspect of human social life and interaction, and as such it is a topic that deserves sociological attention.

Similar observations can then also be made about science. The day to day lived experience of science and scientific practice is as much overlain with humour as any other human activity would be. One of the few classic studies on humour within science has therefore looked at this everyday lived experience of scientists and has indeed found much humour within it [Mulkay and Gilbert, 1982]. Humour by scientists also appears in more formal settings, i.e. not merely when they are talking amongst themselves as colleagues and friends, but within science communication exercises, either as a by-product of the science communicator wanting to adopt a more informal style that is enjoyable for the audience, or in the more direct use of stand-up or other comedy formats as an explicit method of science communication [Bright Club, nd; Bultitude, 2011; Pinto et al. 2013].

The humour in and about science that has been analysed so far appears in most places in formats and situations that are clearly demarcated as humorous or potentially humorous spaces, be it in the conversations between colleagues, cartoons pinned to office and laboratory doors or within science communication; all being reasonably far away from the serious business of science itself. This chapter will instead look at humour in places where there shouldn't be any: serious scientific publications. That humour can appear even within the strict and formal settings of serious scientific publishing is on the surface surprising, but also demonstrates aspects on the nature of humour itself, and the rhetorical and social functions it can serve.

The next section will take a brief overview on the sociology of humour applied to science as well as rhetorical studies on scientific writing and in the process develop some thoughts on the development of scientific seriousness and its rhetorical functions. I will then present three case studies of humour within serious science publications and in the final section present an argument on how the analysis of serious scientific humour can help understand not just the rhetorical uses of humour in science more generally, but also how scientific identities are forged and maintained through humour.

10.2 Humour and seriousness in science

10.2.1 Humour

Mulkay and Gilbert's [1982] analysis of the humour used by the scientists they interviewed for their wider project [Gilbert and Mulkay 1984], has set the stage for more sociological analyses of science humour, which unfortunately never quite materialised. The humour Mulkay and Gilbert saw within their study was analysed as arising through an essential incongruity, following one of the major traditional theories on humour which argues that we find things funny when they juxtapose two incongruous concepts (for more detail see Morreall [1983]; Billig [2005]). The incongruity Mulkay and Gilbert saw was the juxtaposition of two interpretative repertoires which they demonstrated within the general talk of their scientists as part of their wider study; these two repertoires are broadly a formal empiricist repertoire that takes on an empiricist vocabulary and worldview and is typically found within scientific publications, and an informal contingent repertoire that draws on everyday vocabulary of human agency within the scientific work, and its many contingencies and uncertainties. One of the types of joke that Mulkay and Gilbert use to demonstrate the juxtaposed incongruity in mixing the two repertoires is a list of phrases that contrast what a scientist writes within their professional publications and what he or she really means; these lists of what they call "proto-jokes" were found pinned to laboratory notice boards during their visits interviewing scientists for Mulkay and Gilbert [1984]. For example,

> What he [the scientist] wrote: It has long been known that...
> What he meant: I haven't bothered to look up the reference.
> [Mulkay and Gilbert 1982, p. 593]

Through this switch of repertoires, humour is produced, because it re-imagines the science rhetoric used in serious publication with an often more honest appraisal of the human contingencies of actual scientific work.

While Mulkay and Gilbert's analysis does well in describing the origin of the humour within these types of scientists' jokes, it also leaves out other questions that they did not have the space to address themselves. We can look towards more general social psychology of humour to sketch out the wider sociological and socio-psychological aspects at work within these jokes. As Martin's [2007] review outlines, humour can be seen as an identity marker that shows the joker is part of the desirable social group and simultaneously excludes members of the outgroup. Clear examples of this dynamic are deprecating jokes towards minorities. However even very innocuous jokes can fulfil a similar function, for example when people (even self-deprecatingly) poke fun at the social realities lived through only by the ingroup, as with the example of the scientists' jokes analysed by Mulkay and Gilbert: Understanding these jokes requires at least a familiarity with the lived reality of science, and actually finding them funny probably requires lived experience even more so. Jokes that rely on specialist background knowledge likewise delineate between those who get the joke and those who don't and who are therefore clearly demarcated as outside the target audience of likeminded scientists and people who at least know enough about science. We can make a similar point about the intrinsic social division inherent in adolescent sexual humour, which can easily show up listeners who don't understand the joke as sexually inexperienced and therefore explicitly excluded from the target audience and furthermore become socially stigmatised; for this reason it is a probably very common experience to have laughed as teenagers at sexual jokes we didn't actually understand.

Because humour is pleasurable for most people, there is a wider social perception of humour as an unequivocally "good" thing [Billig 2005], where the possession of a sense of humour is seen as an almost morally desirable attribute, for example in personal ads, and humour is seen as a "social lubricant" that brings people together through shared laughter. However as the preceding argument attempted to show, the social lubricant function can be inclusive within a selected social group, but fulfil the opposite function outside it; humour is morally more ambiguous than we usually tend to think. The perceived positiveness about humour extends to the point, as Billig [2001] argues, that unambiguously bad instances of humour, such as racist jokes, tend to be dismissed as "not proper humour" in order for analysts to be able to maintain a "humour-positive" outlook. This, as Billig shows, is a post-hoc rationalisation that has relatively little empirical evidence or logical reasoning behind it.

Using Bourdieu's [1984] conceptualisations of status distinction in the appreciation of art, Friedman [2011] argues that appreciation of humour rests on strategies of social distinction and is therefore a public signal of an individual's cultural capital. Appreciation of a particular kind of comedy as an art form is – similar to the appreciation of abstract vs. traditional painting or avant-garde vs. popular music – a signal of cultural taste and therefore a signal of social belonging. Widening this out to informal humour occurring not just as part of a clearly delineated comedy performance, just as the audience signals their cultural tastes through the appreciation or otherwise of particular types of comedy, so do the jokers themselves signal their social status through the type of humour they engage in.

Humour can break social taboos in relatively safe ways, the habit of adding "only joking" to potentially offensive remarks is a tactic that allows socially taboo topics and opinions to be addressed while not breaking any taboos by itself (though depending on the seriousness of the taboos such as racist, sexist and similar expressions, humour may not always be seen as a valid defence). This gives humour a further social function of allowing forbidden conversations to exist, and explains a variety of both broadly positive applications (such as political satire poking fun at the elite) as well as negative (such as the aforementioned racist jokes). This subversive element of political satire humour however has two sides. While humour may allow political and social dissent to be expressed when other more direct ways of expression are curtailed, it can also become a released pressure valve that makes people believe at least some action has been taken and therefore become complacent about actually getting involved in more affirmative (and effective) political action. The failed subversiveness of comedy can be seen in the *carnivalesque* famously analysed by Bakhtin [1984], or in Mulkay's [1988] review of the political sitcom *Yes, Minister* as ultimately failing to undermine the political class it satirises. While political satire may rarely be on the agenda in humorous scientific publications, a wider subversive voice might well be present, when for example surreptitiously criticising other scientists in a way that would be impossible to do more directly. Humour can however also be used to police established social structures and norms, for example through poking fun at members of the outgroup, or policing deviant behaviour of ingroup members [Martin 2007, p. 119]. As an early seminal study by Coser [1960] found, humour served to reinforce hierarchical status positions among the staff members of the hospital she studied (see also Sayre [2001]).

The social function of humour, like any other human activity, is therefore complex and multifarious. It can serve to break barriers within certain contexts, but can also build them or strengthen others within different contexts. It can be used to undermine and satirise but also to protect the status quo and reinforce hierarchies. By challenging accepted norms of scientific writing, humour within the scientific literature can do all of these things, it can potentially break barriers by showing a human side of science, but simultaneously erect new ones by making clear who the insiders and outsiders are. It can challenge power structures through the deliberate breaking of norms while at the same time reinforcing the power structures or at least showing clearly where the fault-lines of power within science lie. These two aspects of science humour will be explored through three examples below, after a brief introduction to provide the rhetorical background within which humour in science happens and which therefore outlines which rules and norms exactly are broken by humour, and why they are traditionally deemed so important.

10.2.2 Seriousness

Scientific rhetoric in journals is notoriously inflexible and rigid. In contrast to other types of writing by scientists (popular articles and books, some textbooks etc.), and academic writing in the humanities and the more qualitative social sciences, scientific rhetoric often features a passive voice, no personal pronouns and an overall emphasis on preciseness over verbal flourish. Rather than representing a lack of rhetorical technique however, this scientific writing style can itself be analysed as a rhetorical tool to increase the persuasiveness of the argument being made: by eliminating the authors' agency through the use of passive voice and the avoidance of the personal pronoun, the writing demonstrates the unchangeable character of nature: The scientific advance described is not the result of the personal ingenuity of the scientists, but is the result of nature being uncovered and therefore the results as described should be the same regardless of who wrote the paper and performed the experiments.

Likewise, preciseness in expression is a rhetorical tool that emphasises the serious character of the issues being discussed in the paper; the scientific article is not meant as such to persuade the reader – that is the job of the results described within the paper – but to persuade the reader that a serious attempt at disinterestedness has been made, with disinterestedness having been identified by early sociologists of science as one of the most important scientific values [Merton 1973]. Furthermore the reader needs to be able to replicate the results from reading the article, and for that to work, the procedures need to be described in as precise a manner as possible, and the text itself cannot have any playful double-meanings, puns or verbal flourishes. These are finally also restricted by the limited amount of words that most scientific publications allow per article.

These issues surrounding scientific rhetoric are a mixture of necessity and social convention. The preciseness of description is very much a necessary feature of scientific writing, in that it fulfils the function of replicability of the results – although in practice most results are never even attempted to be replicated and even when an attempt is made, often rely on the accumulated tacit knowledge of the relevant scientific community [Collins 1974] that is not transmitted in the article. Nevertheless, preciseness is not something that can be done away with in scientific writing. On the other hand, the passive writing style, the rejection of personal pronouns and the lack of verbal flourish are conventions that we can imagine science writing to abandon and still retain its original purpose. What would be lost if these were abandoned, is the appearance of professionalism, the often acknowledged white lie that nature is all that determines a scientific result and, for the individual author, the cultural capital that accrues with proper adherence to scientific writing standards: an aspiring postgraduate or postdoctoral student will not as a rule want to jeopardise their career prospects by deliberately flouting community norms and accepted standards of professionalism.

It is in this context that deliberate transgressions of these norms through the use of humour become interesting, as Harmon and Gross [2007, p. 221] point out, deviations from the norm can be used to illuminate them. This case studies presented here will attempt to analyse humour in scientific publications by keeping an eye on the cultural value of being relatively humourless and what potential benefits might accrue the author through deliberate transgression of this cultural norm. This is not to say that benefits necessarily do accrue, or that this was one of the reasons the humour had been inserted. On the contrary, an author might have deliberately inserted humour in their publication in the knowledge or expectation of disapproval from their professional peers but with the decision that they simply do not care – this however, as I will argue, is a more likely reaction when the risks of such behaviour are minimised, i.e. it is an avenue open to established scientists rather than those starting their careers.

10.3 Humour in science: three examples

The interwoven themes of humour serving a complex set of social functions and the rhetorical nature of scientific discourse in serious publications therefore provide for an interesting case for further analysis. The analysis of the three cases I will attempt here will serve to hopefully highlight that humour within science is more than just a light-hearted attempt to liven up scientific writing and break barriers by showing a human side to science, but instead is a rhetorical technique that actually may end up highlighting and strengthening deep divisions within science between those that have a considerable amount of institutional habitus and those that don't, and/or as a strategy for saying things that would otherwise be censored, thus showing how in different contexts humour can serve to reinforce social norms and structures and in other contexts can challenge them.

The first example will highlight the nature of power hierarchy within scientific institutions by focussing on the authorship of humorous papers and the differential in risk undertaken depending on the author's seniority. The second will highlight the ability of humour to shut out unconventional and norm-breaking voices and thus reinforce orthodoxy and police deviant behaviour. The third finally will look at the use of humour by outsiders (authors so far on the outside that they have little to lose by engaging in this strategy) challenging the established authorities of the discipline.

10.3.1 The Alpha, Beta, Gamma paper

In 1948 a famous physicist, George Gamow, together with his graduate student Ralph Alpher published an eventually fairly influential paper on "the origin of chemical elements" - Gamow, already notorious then for his sense of humour, added the name of another famous physicist, Hans Bethe, as an "author in absentia" to the submission so that the list of authors read "Alpher, Bethe, Gamow" as a pun on the first three letters of the Greek alphabet, which in turn allowed Gamow and Alpher to name their explanation the "alpha-beta-gamma" theory of the origin of chemical elements [Gross and Harmon 2007, pp. 228-229]. Having been sent the paper by the journal editor, Bethe approved the publication (having struck out the "in absentia" qualification), but otherwise had no input on the paper. Gamow afterwards liked to joke that he tried to persuade another of his graduate students to change his name to Delter [Gamow, 2004].

Briefly discussed by Gross and Harmon [2007] as an example of an amusing deviation from the norms of scientific publishing, this paper also became influential for its content rather than the pun and a springboard for Alpher's subsequent career, so it was hardly a detrimental move for the career of the graduate student involved here. Nevertheless, it also highlights the power difference between the supervisor and the graduate student – it was Gamow's idea to perform the pun, and particularly without having consulted Bethe before sending it out for publication, the move could have potentially detrimental effects on the relationships between the physicists involved had Bethe not shown to have possessed a sense of humour here. Similarly, should the journal editor, or the wider academic readership have reacted negatively to the joke, serious repercussions to the authors could have resulted.

For Gamow, with an already secured academic position and being a widely regarded authority, the risks of potentially souring relationships with other influential physicists would have been there, but not career-threatening, even though, as Alpher recalled, Gamow's sense of humour did indeed not always find appreciation within the wider community, which in turn influenced the choice of which journal to send the paper to [Harwit 1983].

For a PhD student just embarking on a career, the risks must have seemed rather different, although association with Gamow may have lessened them somewhat. While I found no particular indication that Alpher was unhappy about the joke, it nevertheless highlights the fact that norms are more easily broken when the potential adverse consequences are secured against. This first example then illustrates how it is easier for big names to make jokes in their professional writings: the risks to reputation and career in transgressing widely accepted social norms are naturally higher for people starting out than for famous and tenured faculty. While it would be instructive to perform a robust quantitative analysis of scientific papers to find out whether jokes are indeed made more often by senior scientists, the underlying unequally distributed risks are clear.

Publishing jokes in the scientific literature is not simply a matter of confidence of senior authors, but conceivably also one of seniority influencing editorial decisions. An editor will be less likely to dismiss a jokey paper sent in by an established senior figure than by a postdoc they have never heard of. The fact that editorial and even peer reviewer decisions are influenced by the seniority of authorship is a well-known fact of scientific publishing (e.g. Hojat et al. [2003]), and is one of the factors that bring together the so-called "Matthew effect" in the sociology of science [Merton 1973], which will be revisited below.

This effect of seniority making jokes easier also conceivably works on an institutional rather than just individual level. The established and highly ranked *British Medical Journal* for example publishes an annual

Christmas edition devoted to spoof and non-serious papers[1] – examples include: "Were James Bond's drinks shaken because of alcohol induced tremor?" [Johnson et al. 2013]. Emulating this tradition, one might reasonably assume, would be impossible for less established or newer medical journals.

Jokes in serious publications therefore are, among other things, clear signs of the institutional social capital of the joker, and similar to the results in socio-psychological studies of humour as reviewed by Martin [2007], show humour as a potential rhetorical strategy available to those higher up the hierarchy. Not only are less established scientists less likely to want to make jokes in their publications, they are possibly also less likely to get them past the editorial process. The effect that seniority has in the ability to crack jokes in professional contexts, extends to other arenas as well, such as, as mentioned above, professional dynamics in hospitals [Coser, 1960; Sayre, 2001]. Joking in professional context then opens up wider questions about the social effect of humour. As potentially with the Alpher et al. paper, junior collaborators may not be in a powerful enough institutional position to influence the humour of the senior partners, and they may then find themselves manoeuvred into potentially career harming situations. Rather than just showing a light-hearted side of science, the fact that it still breaks very clear and established social norms, humour in science is a complex interaction between social status, risk-taking behaviour and overt signalling of the two.

While this example was used to show that seniority affects the ability as well as the social desirability of joking in professional contexts, the joke here in itself was rather innocuous and had no obvious target at whom the humour was aimed, and therefore did not as such fulfil a higher rhetorical function of policing disciplinary behaviour. The following examples will show how that can be done, first by insiders to the discipline disciplining deviant behaviour and the second by outsiders disciplining a central authority.

---

[1] http://www.bmj.com/about-bmj/resources-authors/article-types/christmas -issue (accessed 15.05.2015)

10.3.2 The best abstract ever!

While the first example is somewhat dated, the second one came about more recently as an answer to a sensational claim in 2011 by scientists working on the OPERA project who claimed to have detected neutrinos travelling faster than light[2]. While the purported discovery made plenty of excited newspaper headlines, professional scientists working in high energy physics were more skeptical. The physicist and science communicator Jim Al-Khalili for example stated that if the claims turn out to be true, he would eat his boxer shorts on live television[3]. Similar to previous sensationalist discoveries such as the famous cold fusion case [Lewenstein 1995], the scientists at OPERA opened themselves up to a lot of criticism, disdain and disapproval from the wider physics community, not only because it was argued that if true, we would have to rip up our physics textbooks, but also because the claim was made via press-release rather than the usual avenue of peer reviewed scientific journal publishing. There were correspondingly many people who drew parallels between the OPERA results and previous "junk" or "pseudo-scientific" discoveries. This dismissive attitude (regardless of its factual merits) is crucial here, as it cast the OPERA scientists as potential figures of scorn and derision.

Several rebuttals were published in the months after the announcement. One of them, titled "Can apparent superluminal neutrino speeds be explained as a quantum weak measurement?" [Berry et al. 2011] featured only a two word abstract: "probably not". The paper made the round in informal emails between physicists and beyond, and was featured on the igNobel prize website as "the best abstract ever"[4]. The joke abstract signalled several things which could be appreciated even without special knowledge of the contents of the rest of the paper: the curt dismissal of the OPERA claims in the abstract made the dismissive attitude of the wider physics community appear to be obvious even to

---

[2] http://press.web.cern.ch/press-releases/2011/09/opera-experiment-reports-anomaly-flight-time-neutrinos-cern-gran-sasso (accessed 15.05.2015)

[3] http://news.bbc.co.uk/1/hi/programmes/newsnight/9598802.stm (accessed 15.05.2015)

[4] http://www.improbable.com/2011/10/14/ig-nobel-winner-writes-best-abstract-ever/ (accessed 15.05.2015)

those outside the specialism, a clear rhetorical strategy that helped to undermine their results. It also clearly signalled that the claims are not to be taken seriously simply through the fact of being humorous.

The effect was no doubt amplified, at least within the specialist community, by the fact that the lead author of the rebuttal is a well-established and regarded scientist, with a fellowship of the Royal Society.

The fact that it is an established senior scientist within the discipline who makes this humorous transgression is then a fairly clearly an example where humour is being used as a way of denigrating deviant behaviour within the discipline. The humour is being used as a device to make the deviant scientists appear incompetent and unscientific. The transgression of the norms of stylised scientific writing – while by itself a clear violation of accepted practice – in fact serves to strengthen the orthodoxy and established ways of doing things. The satirised scientists themselves broke arguably more serious norms by rushing to declare results that in all probability were erroneous, and not only that, they did this publicly through press release rather than peer-reviewed scientific publishing.

10.3.3 The complex dynamics of wishful thinking

The act of norm and taboo-breaking itself has been pointed towards as one of the wider social functions of humour. Not only has a norm been violated producing the humour through incongruity, but also the clear signalling of the violation as humorous actually helps in defusing the situation again: Other potential transgressions from the norms of academic writing (for example offensive language and swearing) would be regarded as more serious. There are even fewer potential examples of this that I could find and when they do occur, they can often be constructed or re-interpreted as an attempt at humour. Correia's [2013] rather ill-tempered response to Jared Diamond's popular science books (entitled "F**k Jared Diamond", asterisks, thankfully, in the original) would be an example here: From the contents it is clear that the feelings of the author towards Diamond as reflected by the title are sincerely held and therefore it is not merely an attempt at jocularity. Nevertheless the title appears to be an attempt at a humorous transgression of the rules of academic writing (though in this case, social scientific writing where traditionally the style rules are rather more relaxed).

The rhetorical defence of "I was only joking" is a method for people to express taboo feelings and subjects without attracting the moral disapproval they might have caused had the issues been raised in a serious manner, and/or to test the boundaries within a group of where the taboos lie. Sexually explicit jokes, racist jokes, political satire in repressive regimes, all share the feature that being labelled as jokes makes them socially more acceptable. In the second example above, the joke works rather clearly on that level – what the authors signal through their curt abstract is that they think the scientists at OPERA were violating serious scientific norms, yet by formulating it as a joke they were able to get the message across fairly clearly to anyone without actually saying so and thus breaking even stronger social taboos than the ones governing humour in science publications that usually would prohibit an author to directly state that other scientists are incompetent. This tactic is even more pertinent when outsiders argue against the received wisdom of the discipline, where it would be even more impossible to make straight accusations of incompetence.

The art of the humorous put-down was demonstrated in a paper by Brown et al. [2013] which convincingly argued that a recent but already seminal theory in the field of positive psychology [Posada and Frederickson 2005] rested on a few fundamental misunderstandings of basic mathematics and physics. Losada and Frederickson used differential equations developed within fluid dynamics to model human emotions and arrived at a – with hindsight – suspiciously precise "positivity ratio" of 2.9013 positive to negative emotions. Being above that tipping point, they argued, allows people to flourish, while being below that ratio, lets people languish. The positivity ratio has in a relatively short time become a highly cited and influential theory, and became one of the cornerstones of positive psychology.

Needless to say, Brown et al.'s paper, convincingly demonstrating that the positivity ratio is fundamentally flawed, created quite a stir within positive psychology because it by extension also argued that a whole academic field has been negligent in allowing itself to be partially built on fundamentally flawed reasoning. While Brown et al. did not write their paper directly suggesting that Losada and Frederickson were incompetent, the general tone was clearly and harshly critical of their work, though still staying within the norms of accepted practice when criticising the work of other academics. However, the more humorous analogies they inserted into the paper were clearly suggestive of exactly what they thought, starting with the title itself: "The complex dynamics of wishful thinking", which parodies the title and one of the central claims of Losada and Frederickson [2005], i.e. that "human flourishing" can be modelled through the complex dynamics developed in physics to describe fluid systems (Losada and Frederickson [2005] is titled: "Positive affect and the complex dynamics of human flourishing")[5]. A slightly longer quote from Brown et al. demonstrates the not so subtle disparaging humour they employed to criticise Losada and Frederickson's reasoning:

> They appear to assert that the predictive use of differential equations abstracted from a domain of the natural sciences to describe human interactions can be justified on the basis of the linguistic similarity between elements of the technical vocabulary of that scientific domain and the adjectives used metaphorically by a particular observer to describe those human interactions. If true, this would have remarkable implications for the social sciences. One could describe a team's interactions as "sparky" and confidently predict that their emotions would be subject to the same laws that govern the dielectric breakdown of air under the influence of an electric field. Alternatively, the interactions of a team of researchers whose journal articles are characterized by "smoke and mirrors" could be modeled using the physics of airborne particulate combustion residues, combined in some way with classical optics. [Brown et al. 2013]

[5] Their original proposed title "The Complex Dynamics of an Intellectual Imposture" was rejected by the editor as implying deliberate fraud.

Where the "smoke and mirrors" here is a veiled but easily spotted reference to what they think of Losada and Frederickson's output.
Thus, through humour, the authors were able to make pretty serious points without breaking higher social taboos. The same authors were publicly more candid about their feelings in more informal settings, such as newspaper interviews[6], so this was not a matter of them wanting to spare anyone's feelings, but one of being able to put their criticisms into a format that it could be published in a scientific journal.

Above I have described this paper as written by outsiders criticising an established figure within the discipline, and the story of the genesis of this paper is actually quite interesting. Brown, the lead author, was an MSc student on a positive psychology course at the time who – without claiming any particular expertise in mathematics – had become suspicious of the reasoning behind the core readings handed out to him at a lecture. Not feeling that he had the sufficient knowledge of mathematics to formulate his suspicions, he contacted the physicist Alan Sokal – famous for his spoof paper that helped spark the so-called Science Wars of the mid-1990s [Sokal 1996] – for help, as well as some established but potentially sympathetic psychologists who encouraged Brown to publish. The eventual team of authors thus featured a graduate student, a physicist and a psychologist from outside the subfield of positive psychology. Brown was clearly not in an institutionally powerful academic position and although the two other authors are, they also are from different or at best related disciplines than the one that has been criticised.

10.4 To those that have, shall be given; to those that joke shall be listened to

In the examples above, as well as within most other humorous insertions into the scientific literature, the humour itself works through the incongruous juxtaposition of non-serious language into a serious piece of writing, much like the switch between the empiricist and contingent

[6] http://www.theguardian.com/science/2014/jan/19/mathematics-of-happiness-debunked-nick-brown (accessed 15.05.2015)

repertoires analysed by Gilbert and Mulkay [1982]. This in turn makes phrases and puns that would otherwise possibly not be particularly funny (depending on one's sense of humour) appear even more humorous. It is thus somewhat paradoxically the non-humorous nature of scientific writing that is partly the source of the humour itself – the incongruous transgression of the otherwise very strong norms of stylised writing in science.

While incongruity is one of the mechanisms through which the humour works, on a further level it also relies on background knowledge of the intended audience. Alpher, Bethe, Gamow may be a pun that would be recognisable for an audience of scientists who use the Greek alphabet routinely in their work, it might not appear obvious to people outside. Likewise, the transgression committed by Berry et al. in providing a two-word abstract would only be recognisable by readers familiar with the norms that are being transgressed – other scientists again, and those familiar with scientific writing practices. Through this, the jokes here also work through the appeal to an insider audience; they are jokes that not everybody will be able to appreciate, but primarily only those within the jokers' intended audience. The effect of humour as a "social lubricant" arises here, because it binds the community together in the appreciation of a shared joke, though it simultaneously also excludes those on the outside (and of course, those who are the butt of the joke). Unlike some other humorous situations, this is here a broadly positive effect as for scientific papers there is almost no audience beyond those who would understand and appreciate the joke. Nevertheless it also limits the potential reach of the joke as well.

The appearance of humour in serious scientific publications violates clear socially established norms and values within the wider scientific community. This is one reason why it doesn't happen particularly often, but on the few occasions that it does, it demonstrates some interesting facts about the social capital and institutional status of the authors as well as potentially their rhetorical strategy in capitalising on it. Jokey papers bring themselves to a wider audience, by being shared more readily between scientists in the discipline and even beyond. The fact that this type of strategy – because of the strong norms discouraging it – is open mostly only to senior and established scientists enhances one of the outcomes of Merton's Matthew effect in that it is easier for scientists with an already established reputation not only to spread their research further than non-established scientists, but also in being more successful in demonstrating to the wider world beyond their immediate circle, that they have a sense of humour. As Billig [2005] has argued, having a sense of humour has over the past century, at least within Anglophone societies, become an almost moral quality, the demonstrated possession of which makes people appear more sympathetic and human. The effect then of humour in serious science is therefore not to merely to accentuate the Matthew effect as a furtherance of scientific ideas as Merton envisaged, but also to increase the social and cultural capital of the scientists beyond their specialism by making them appear more sympathetic than scientists further down the hacking order who are not as able to publish funny papers.

The specific uses of humour as a rhetorical device to dismiss others, as in the second example, enhances this effect; the author not only manages to portray themselves in a positive light, but simultaneously damages the social credibility of those they satirise. While this may well be deserved in the example given, it also demonstrates a darker side of humour, which has been analysed more extensively through the clear-cut examples of racial or nationalist humour [Billig 2001]. But it applies more generally to any humour that has a clearly defined target; some traditional philosophical analyses of humour such as those of Plato and Hobbes (subsequently labelled "superiority" theories on humour) saw humour to be mostly about making the joke-teller and their ingroup audience feel superior by deprecating others. The unique effect of humour is that while it is a clear violation of scientific writing norms, it is also a violation that is in some sense permissible.

10.5 The fine balance of humour and seriousness in the scientific literature: Have we reached peak funny?

There is an ambivalence in the use of humour within serious scientific publishing in that on the one side it is a violation of established norms and frowned upon as such, but on the other, humour carries a wider cultural cachet that certainly within contemporary Anglophone society is seen as a clear, almost morally, good attribute. Within the general corpus of scientific literature there are thus two culturally contingent forces pulling into opposite directions; a traditional one towards an impersonal, objective style and another one into a light-hearted and humorous one. Despite the risks, the benefits for the individual author to publish humorous material in the serious literature are several, including wider recognition not just as a scientist and also as potentially a scientist of secure enough standing to be able to do this, but also as a person being seen to possess a personal attribute that is morally valued within contemporary society. Other benefits include being able to make commentary on matters in a way that would be impossible outside a humorous context, this rhetorically strengthens the scientists' position and manages to discredit those of whom they disapprove and thus work to strengthen existing social boundaries as well as setting up new ones. Even at its most innocuous, i.e. when there are no clear targets of the humour as in my first example, science humour still relies on an implicit division between insiders and outsiders, thus strengthening the social identity of the relevant scientific communities, but also potentially erecting boundaries beyond those who are expected to understand and appreciate the joke.

Yet as a wider strategy for writing scientific articles, humour is still being discouraged, and probably rightly so, as the epistemological status of science does seem to depend on it presenting a serious face to the world, and if the norms were dropped and suddenly anything goes, then science would potentially lose much needed epistemic credibility among the public – therefore what seems to be beneficial for the individual scientist is potentially bad for the activity of science as a whole. Furthermore, as the humour in most cases relies on the incongruity of jokes appearing in serious places, too much joking would dilute the impact of those occasional jokes that do appear, both because it would render them less funny as the unexpected incongruity between in the switch between interpretative repertoires becomes less and less unexpected, as well as because it would no longer be much of a novelty and therefore won't make the author stand out as much. We should therefore not expect to see the barriers torn down completely, and humorous writing to remain the exception within scientific publishing.

Therefore, rather than humour being just a way of scientists showing off their human side, I have argued that it can fulfil several other social and rhetorical functions and/or unintended side-effects that are in some ways less benign.

References

Bakhtin, M. [1984] *Rabelais and his world* (Indiana University Press, Bloomington).

Berry, M. V., Brunner, N., Popescu, S., & Shukla, P. [2011] "Can apparent superluminal neutrino speeds be explained as a quantum weak measurement?". *Journal of Physics A: Mathematical and Theoretical*, 49, 492001.

Billig, M. [2001] "Humour and hatred: The racist jokes of the Ku Klux Klan", *Discourse & Society* 12(3), 267-289.

Billig, M. [2005] *Laughter and ridicule: Towards a social critique of humour.* (Sage, London).

Bourdieu, P. [1984] *Distinction: A social critique of the judgement of taste* (Harvard University Press, Cambridge, MA).

Bright Club, [no date] *About Bright Club.* http://brightclub.wordpress.com/ (accessed 7.05.2015).

Brown, N. J., Sokal, A. D., & Friedman, H. L. [2013] "The complex dynamics of wishful thinking: The critical positivity ratio", *American Psychologist* 68(9), 801-813.

Bultitude K [2011] "The why and how of science communication" in *Science Communication,* ed. Rosulek, P. (European Commission, Pilsen).

Collins, H. M. [1974] "The TEA set: Tacit knowledge and scientific networks", *Social Studies of Science*, 4(2), 165-185.

Correia, D. [2013] "F** k Jared Diamond", *Capitalism Nature Socialism* 24(4), 1-6.

Coser, Rose Laub [1960] "Laughter among colleagues: A study of the functions of humor among the staff of a mental hospital", *Psychiatry* 23, 81-95.

Fredrickson, B. L., & Losada, M.F. [2005] "Positive affect and the complex dynamics of human flourishing", *American Psychologist* 60(7), 678–686.

Friedman, S. [2011] "The cultural currency of a 'good' sense of humour: British comedy and new forms of distinction", *The British Journal of Sociology* 62(2), 347-370.

Gamow, G. [2004] (1952) *The Creation of The Universe* (Dover Publications).

Gilbert, G. N. & Mulkay, M. [1984] *Opening Pandora's box: A sociological analysis of scientists' discourse.* (Cambridge University Press, Cambridge).

Hojat, M., Gonnella, J. S., & Caelleigh, A. S. [2003] "Impartial judgment by the "gatekeepers" of science: fallibility and accountability in the peer review process", *Advances in Health Sciences Education* 8(1), 75-96.

Harmon, J. E., & Gross, A. G. (eds.) [2007] *The scientific literature: A guided tour* (University of Chicago Press, Chicago).

Harwit, M. [1983] *Interview of Ralph Alpher by Martin Harwit on August 11, 1983*, Niels Bohr Library & Archives, American Institute of Physics, College Park, MD USA, http://www.aip.org/history/ohilist/3014_1.html (accessed 7.05.2015).

Johnson, G., Guha, I. N., & Davies, P. [2013] "Were James Bond's drinks shaken because of alcohol induced tremor?", *BMJ: British Medical Journal*, 347.

Lewenstein, B. V. [1995] "From fax to facts: Communication in the cold fusion saga", *Social Studies of Science* 25(3), 403-436.

Martin, R. A. [2010] *The psychology of humor: An integrative approach* (Elsevier Academic Press, London).

Merton, R. K. [1973] *The sociology of science: Theoretical and empirical investigations* (University of Chicago Press, Chicago).

Morreall, J. [1983] *Taking laughter seriously* (State University of New York Press, Albany).

Mulkay, M. [1988] *On Humour: Its Nature and Place in Modern Society* (Polity Press, Cambridge).

Mulkay, M. & Gilbert, G. N. [1982] "Joking apart: some recommendations concerning the analysis of scientific culture", *Social Studies of Science* 12(4), 585-613.

Pinto, B., Marçal, D., & Vaz, S. G. [2013] "Communicating through humour: A project of stand-up comedy about science", *Public Understanding of Science*, published online before print, doi: 0963662513511175.

Sayre, J. [2001] "The use of aberrant medical humour by psychiatric unit staff", *Issues in Mental Health Nursing* 22(7), 669-689.

Sokal, A. D. [1996] "Transgressing the boundaries: Toward a transformative hermeneutics of quantum gravity", *Social Text* 46(47), 217-252.

11

From Science to Fiction

Nigel Sanitt

A small number of scientists occasionally cross the divide between science and the arts and write novels. Many such novels are in the science fiction genre, but some scientists delve into other types of fiction writing. For every scientist who publishes a novel, there are a greater number that dally with the idea of writing fiction, but for various reasons never quite make it. This article is directed at this group with the hope that it might encourage some to take the leap. How different is fiction writing from science writing? How does one find a publisher? What do you tell your friends and colleagues? The present writer is a scientist who has made the leap, having written two novels[1], under the pen name of *Norman Stanton*.

11.1 Introduction

You want to write a debut novel, have it published and be a successful author, but to date you have only written scientific papers.

The good news is you can – the bad news is you can't - I'll explain this conundrum later.

Scientists generally don't write novels. Even if they have a go, they usually come up with science fiction stories. There are, of course, notable examples: Isaac Asimov (1920-1992) a professor of biochemistry famous for his *Foundation* series. H G Wells (1866-1946), the author of *The Time machine* and many other novels, who studied Biology under T H Huxley at Imperial College London.

Moving away from science fiction, examples of scientists who have crossed the divide are the novelist C. P. Snow (1905-1980) who had a

---

[1] Perfect Nightmare (Thames River Press, London, 2013) and The Literary Circle or the case of the arrogant anagram (Pantaneto Press, Luton, 2014).

PhD in Physics, the writer Primo Levi (1919-1987) who was a chemist, and in more recent times, Alan Lightman, Professor of theoretical Physics and of Humanities at MIT and Paul McEuen, Professor of Physics and Humanities at Cornell.

Paul McEuen, when asked how he became a thriller writer, replied that he felt he had three choices when he was going through a midlife crisis: “Buy a sports car, become a dean or write a novel.”
I don’t know whether he achieved the first two of these but he did write novels.

One of the authors I referred to – C. P. Snow, wrote about the divide between science and the arts in a famous article based on a talk he gave at Cambridge University in 1959, entitled *The Two Cultures* [Snow 2012 (1959)]. I’m sure Snow was influenced by the University Library at Cambridge. Built in the early 1930s, the architect Giles Gilbert Scott (1880-1960) devised an edifice of intimidating proportions. The North wing houses the humanities and the South Wing the sciences. It takes a good ten minute walk from one to the other. With a tower in between (reputedly full of pornography) the building has been described as “designed by Kafka to frighten Freud”. As an aside, Scott also designed the famous red telephone box that graced the streets of London. The intimacy of the telephone box is at odds with the disassociation between science and the humanities in the library.

11.2 Debut Novel

What are the particular problems faced by debut novelists (whether or not they are scientists)?

In order to start a novel, I believe there are at least four things that have to be firmly in place before even the first word can be penned. They are: An idea, a title, an opening page and an ending. The additional requirements an author needs beyond these four are arrogance and weirdness.

Let me explain. I think there is no argument that in order to write a book an author needs an idea – an overarching idea, which reflects the motivation and the passion behind the work. This is how novels start.

Why is a title so important? I do not refer to the actual title of the book but the working title. The difference is that the working title is a pithy summary of the book which helps the author to stay focused. The actual title is often chosen by a publisher (if an author is lucky enough to find one) and is more to do with the marketing of the book.

It takes a long time to write a novel and it is easy to get lost. A title is something for a writer to hold on to when the going gets rough – which it will.

The opening page is a vital ingredient for any writer. There is nothing more likely to cause anxiety to an author than being faced with a blank page, especially when it is the first page of a book. That is why it is important that the first page is already written in an author's mind so that starting a book is not a problem.

By ending, I do not refer to the last page of a book but how the story ends. It is easy to get lost when writing a book and an *ending* gives an author a road map, so he or she knows where they are going.

I mention arrogance and weirdness because they are characteristics displayed by writers: arrogance, because the author needs to believe that readers really want to read what they have written and weirdness, because without an offbeat way of looking at the world no piece of writing is going to be a *page turner.*

Let us assume that an author has the four requirements in order to start a book and that he or she is sufficiently arrogant and weird to see the work to completion, with an interesting narrative. What problems will be faced, especially by a debut novelist?

First, what is described as *Pace*: debut novelists tend to move a story on too quickly. Word count is an extremely useful tool as it signals to a writer where they are in a book. The typical novel is around 80,000 words, so if you have covered the whole story in 10,000 words, you need to slow down. This is a particular problem, I believe, for scientists. Scientific writing has to be concise with as little extraneous or irrelevant detail as possible. The opposite is the case for fiction writing – although degenerating into waffle is to be avoided.

Second, there may be a problem of *writer's block* – the terrifying blank page that haunts writers. Each writer confronts this problem in their own way. One method, which I use and which, for me at least, is successful, is not to discuss a work with anyone until the first draft is completed. By not confiding in anyone, even those closest to you, the only outlet is the written page. Even if there is a brief period when the muse is not forthcoming, the build up of tension becomes so great that the words eventually find themselves on the page. This does not, of course, make writers easy to live with!

Third, research for a novel is important. A novel is, of course, one big lie, as otherwise it would be non-fiction, but credibility and internal consistency are vital. If a character takes a plane from New York to Lima, Peru, for example, then it is no good having them arrive in the morning – all the planes from New York to Lima arrive in the evening. If you think things like this are not important, just wait until you receive letters from readers pointing out that you got your plane schedules wrong. A scientific background here is useful as careful research is part and parcel of the craft of science.

Fourth, outstaying your welcome is a common fault for debut novelists. If you are describing a scene you do not have to start your narrative at the beginning and stay to the end. If two characters, for example, are having lunch in a restaurant you do not have to include them arriving at the restaurant, hanging up their coats, being taken to their table – unless these things are germane to your story. Drop in and out at the right times.

Fifth, you do need to set a regular time for writing and a goal of how many words you complete in a given time. This can be quite flexible, some authors have a daily goal, and others monitor their output over a longer period. In my case I try to write a given number of words per fortnight, usually 5,000. Even if one does not succeed in keeping up it is a good discipline to try to monitor one's output.

Sixth, debut novelists often forget about the five senses. In addition to sight and hearing there is taste, smell and touch. In particular, smell is an extremely evocative and basic sense. One cannot use it too frequently, but now and then it can be a powerful device. Needless to say if you are trained in science then writing about your feelings through different senses is an unusual experience.

11.3 Re-writing

I was at a lecture given by the ex politician and novelist Lord Jeffrey Archer who was talking about his life as a writer. He admitted that the average number of re-writes for each of his books is seventeen. Furthermore, he freely admitted that his first drafts were mostly unintelligible. Needless to say re-writing a book is extremely important.

Let us assume that you have finished a first draft of your novel, over which you have laboured for many months maybe even years. How do you start to re-write it?

There are a few general points which are useful in order to advance the process. These are represented by the phrases: big then small, show don't tell, nouns and verbs not adjectives and certainly not adverbs.

Some authors decide after the first draft to make major changes to their narrative. These changes must be incorporated as early as possible – start with the big things then work your way down to the small.

Show don't tell is probably the most important universal rule. If a character is angry the reader does not want to be told that the character is angry, much better to have the character smash a plate on the floor.

What do I mean by nouns and verbs not adjectives and adverbs? This is part of a word list of words to avoid, which an author should consider. There is of course nothing wrong with using an adjective or an adverb, even one which ends in -ly. But in order to colour the narrative and raise the literary standard of the writing it is much better to use the right noun or verb rather than relying on a modifier.

An example would be the verb to walk. We can say that a character walks quickly or slowly, but it is far better to say that the character rushes or saunters – readers appreciate the *mot juste*. By the way, foreign expressions should not be overused – it can appear condescending to the reader.

So what are the words to be avoided? Examples are the usual suspects: very, get, nice, rather, quite, and of course the -ly adverbs: certainly, surely, hopefully and others.

There are two caveats that I must put in here. First, all these rules are meant to be broken. The trick is knowing when you can get away with it and improve the writing as a result – not easy.

As an example take the extremely successful book The Infinite jest by David Foster Wallace. It runs to 1079 pages, has very few paragraphs, and is mostly the mental ramblings of a drug addict. *Time* magazine voted it one of the 100 best English-language novels of the period 1923-2005. Sadly, Wallace committed suicide in 2008.

Second, for many writers, myself included, the above list of words to be avoided are useful in helping the flow of writing for the first draft and can be omitted in the second draft. The last thing you want to do is interrupt your writing stream while you think of a particular word that will fit best. This is very much to do with the personal style of writing of an author, some prefer to take their time and re-write fewer times. The red pen is your friend but maybe not to start with.

Finding the right words should be second nature to scientists where precise language is important.

11.4 Publication

You have written your masterpiece, re-written it until you're sick of the sight of the book, now you want to publish it. There is no easy way to say this, so I'll just come right out with it: Debut novels usually fail unless the author is already famous or for other special reasons.

Let me start to explain this by giving a few examples.

Jane Austen

Jane Austen's first novel was Sense & Sensibility, which was published in 1811. There are a few side facts about the publication of this important book, which are relevant to the present discussion. One peculiarity is that Jane Austen's name did not appear. The book is described in the frontispiece as "By A Lady".

Another interesting fact is that Jane Austen was an extremely shrewd businesswoman. She approached a publisher and offered him a deal. She would pay for the first run of 750 copies to be printed and in addition she would pay the publisher a fee to market and sell the books. The publisher could not turn down this win-win deal. The first run was printed and sold out and resulted in a £140 profit for Austen. The rest is history.

Margaret Mitchell

Up until the mid 1990s, *Gone With The Wind* published in 1936 was considered as Mitchell's first novel. In fact it was almost certainly her third. Her first book Last Laysen, written when she was only fifteen years old, was discovered after her death and published in 1996. It was a New York Times best seller. Her second book, which was about 400 pages long was lost and probably destroyed. Mitchell was a prolific writer. She was a journalist for the Atlanta Journal and wrote prior to *Gone With The Wind* more than 120 feature articles and 80 news stories.

J D Salinger

Catcher in the Rye (published 1951) was Salinger's only novel, but he was a short story writer for the New Yorker magazine from 1940.

Harper Lee

Her debut novel *To Kill a Mocking-Bird* came out in 1960. It won a Pulitzer Prize and has to date sold 40 million copies. This is often cited as an example of a successful debut novel. But it was not the first novel she wrote. Recently, a manuscript of a novel *Go Set a Watchman,* thought to have been lost and written before *a Mocking-Bird,* has come to light.

Lee trained as a lawyer during which time she wrote for a University newspaper. She wrote in her spare time while she was living in New York and her book was taken up by a publisher's editor. The original *Mocking-Bird* manuscript was in the form of a series of stories and Lee, with the editor's help, re-wrote the manuscript over a two-and-a-half-year period into a fully fledged novel.

J K Rowling

Harry Potter and the Philosopher's Stone was published in 1997. Rowling went to Exeter University after rejection from Oxford. She became a bilingual secretary (her Mother was French) and taught English as a foreign language at Oporto, Portugal. She had a literary agent but was turned down by 12 publishers.

Her publisher insisted she dropped her first name (Joanne) and instead use two initials. Since she did not have a middle name, Joanne chose the letter K, hence J.K. Rowling.

Iain Banks

The Wasp Factory (published 1984) was Banks's first success. He had written other novels before The Wasp Factory and his first - The Hungarian Lift Jet, was written at the age of 16. When asked by his publishers whether they could publish his first book he replied: "Over my dead body."

Tragically, Banks has since died but I doubt whether this book will be coming out anytime soon.

11.5 Conclusion

There are plenty of other examples I could give which show that successful debut novels are much rarer than most aspiring authors think. The obvious question that comes to mind is that if first novels rarely make it then how do novelists become successful – and a number of them do.

Also, in my opening paragraph I posit a conundrum that first novels can (and can't) succeed. My answer to these points is that success as a novelist is attainable; you just have to write more than one novel.

Novel writing is not a theoretical enterprise, it is a craft. Or at least the craft aspect of writing is of prime importance. Let me give an analogy with cabinet making. A cabinet maker has to become apprentice to a craftsman and before becoming even a journeyman cabinetmaker must spend many years building, for example, desks. Even a highly successful master craftsman would not want to be reminded of his first attempts to build a desk when he was a newly qualified carpenter.

Book marketing

There is an old saying that you have to be clever to write a book but a genius to sell one. These days publishing a book is the easiest thing to do. With *Print On Demand* any author can publish his or her book for a relatively small sum of money. A successful book, however, needs marketing and that is why generally speaking it is better to find a publisher if you can.

Apart from self-publication the other two options are a literary agent or a publisher. Literary agents are generally, but not exclusively, for authors who are either famous names or who have already been published. So how does one find a publisher?

The answer is: with great difficulty. My first book was picked up by the 38th publisher I approached. When I explained this to another author who had successfully published a number of books, he told me that I was lucky – his first book had 74 rejections before he found a publisher.

One thing which will give an author an edge in finding a publisher is if you can offer two or even three books, rather than just one.

So the solution to my conundrum is if you are not already famous and you don't know a literary agent then write at least two books.

My final point is: what do you tell your friend and colleagues? Because I have written in science, I chose a pen name for myself as a novelist, in order not to confuse my fiction and non-fiction writing. It is perfectly OK to tell your scientific colleagues that you are a novelist – but be prepared for some strange looks!

References

Snow, C. P. [2012] (1959) *The two cultures.* (Cambridge University Press, Cambridge).

Index

A

B

C

D

E

S

T

U

V